Expert Aided Control System Design

Other titles published in this Series:

Parallel Processing for Jet Engine Control
Haydn A. Thompson

Iterative Learning Control for Deterministic Systems
Kevin L. Moore

Parallel Processing in Digital Control
D. Fabian Garcia Nocetti and Peter J. Fleming

Intelligent Seam Tracking for Robotic Welding
Nitin Nayak and Asok Ray

Identification of Multivariable Industrial Processes for Simulation, Diagnosis and Control
Yucai Zhu and Ton Backx

Nonlinear Process Control: Applications of Generic Model Control
Edited by Peter L. Lee

Microcomputer-Based Adaptive Control Applied to Thyristor-Driven D-C Motors
Ulrich Keuchel and Richard M. Stephan

Colin Tebbutt

Expert Aided Control System Design

With 49 Figures

Springer-Verlag
London Berlin Heidelberg New York
Paris Tokyo Hong Kong
Barcelona Budapest

Colin D. Tebbutt
LinkData
P.O. Box 23316
Claremont 7735
South Africa

ISBN 3-540-19894-6 Springer-Verlag Berlin Heidelberg New York
ISBN 0-387-19894-6 Springer-Verlag New York Berlin Heidelberg

British Library Cataloguing in Publication Data
A catalogue record for this book is available from the British Library

Library of Congress Cataloging-in-Publication Data
A catalog record for this book is available from the Library of Congress

Apart from any fair dealing for the purposes of research or private study, or criticism or review, as permitted under the Copyright, Designs and Patents Act 1988, this publication may only be reproduced, stored or transmitted, in any form or by any means, with the prior permission in writing of the publishers, or in the case of reprographic reproduction in accordance with the terms of licences issued by the Copyright Licensing Agency. Enquiries concerning reproduction outside those terms should be sent to the publishers.

© Springer-Verlag London Limited 1994
Printed in Great Britain

The publisher makes no representation, express or implied, with regard to the accuracy of the information contained in this book and cannot accept any legal responsibility or liability for any errors or omissions that may be made.

Typesetting: Camera-ready by authors
Printed by Athenæum Press Ltd, Newcastle upon Tyne
69/3830-543210 Printed on acid-free paper

To Mother

"Whatever you do
work at it with all your heart,
as working for the Lord."
– *Colossians 3:23*

SERIES EDITOR'S FOREWORD

The series *Advances in Industrial Control* aims to report and encourage technology transfer in control engineering. The rapid development of control technology impacts all areas of the control discipline. New theory, new controllers, actuators, sensors, new industrial processes, computing methods, new applications, new philosophies, ... new challenges. Much of this development work resides in industrial reports, feasibility study papers and the reports of advanced collaborative projects. The series offers an opportunity for researchers to present an extended exposition of such new work in all aspects of industrial control for wider and rapid dissemination.

Dr Colin Tebbutt has contributed a very interesting volume on the use of expert systems in the control design process. All the research, design, development and testing aspects of creating the expert system are covered in the volume. The expert software has modules to help the novice or expert control designer formulate and refine the controller design specification and uses a recent control design approach to synthesize the controllers. The volume reports both successes and identifies the shortcomings in the expert technological approach. The volume fully documents the use of the expert control design software by novice and experienced users. The very nature of the development and its use should make this volume of interest to a wide range of readers from industrial engineers wishing to speed the design process to those in academia who might be interested in the educational potential of the work for teaching control engineering.

M. J. Grimble and M. A. Johnson
Industrial Control Centre,
University of Strathclyde,
Scotland, U.K.

PREFACE

Successful multivariable control system design demands knowledge, skill and creativity of the designer; without all three of these qualities the design is unlikely to be excellent. The allure of artificial intelligence is the prospect that much of the knowledge and some of the skill of the designer may be captured into an intelligent design tool, leaving the designer free to concentrate more on the creativity aspects of the design. The objective of the research described in this volume has to investigate the contribution which artificial intelligence could make to this design process. More specifically, the investigation strove to develop strategies which could both improve the efficiency of experienced designers, and assist and guide novice designers, and to implement, demonstrate and evaluate these techniques. While existing CACSD tools allow the designer to analyse a control system with a particular controller, in general they do not provide the user with any guidance as to how the design could be improved. Providing this guidance has been an explicit goal for this research.

An intelligent, interactive, control system design tool has been developed to fulfil these goals. The design tool comprises two main components; an expert system on the upper level, and a powerful CACSD package on the lower level. The task of the expert system is to guide the designer in using the facilities provided by the underlying CACSD package. Unlike other expert systems, the user is also assisted in formulating and refining a comprehensive and achievable design specification, and in dealing with conflicts which may arise within this specification. The assistance is aimed at both novice and experienced designers.

The CACSD package includes a synthesis program which attempts to find a controller that satisfies the design specification. The synthesis program is based upon a recent factorization theory approach, where the linear multivariable control system design problem is translated into, and solved as, a quadratic programming problem. Techniques which significantly improve the time and space efficiency of this method have been developed, making it practical to solve substantial multivariable design problems using only a microcomputer.

Initially a single variable version of the design system SV-CXS was built and evaluated as a feasibility study for MV-CXS, the full multivariable design system. Although this early system had several shortcomings, and illustrated many weaknesses in the expert system, most of these could be overcome, and in retrospect the study made no small contribution to the success of the multivariable version.

The intention of this volume is to describe the research conducted and techniques developed in detail. The reader is expected to have an understanding of the broad principles of control system design, including an appreciation for the physical meaning of closed loop responses (time and frequency domain), and of the use of transfer functions. Some knowledge of linear albegra is required, particularly in chapter 2; however, the important concepts discussed can be grasped without any advanced mathematical abilities.

Chapter 1 begins with an introduction to the application of artificial intelligence techniques to control system design. The chapter also includes a discussion on the reasons for selecting an expert system approach, and provides a brief overview of the MV-CXS design system.

Chapter 2 and 3 deal with the multivariable CACSD package in detail, giving details on the theoretical basis and computer implementation. The efficiency of the CACSD design method has been improved substantially relative to that of the standard algorithms. To achieve this, a method for decomposing a large class of multivariable systems into smaller independent sub-problems, and a novel parameterization for approximating the set of stable transfer functions, have been developed. In addition, a compact and computationally efficient representation for storing the linear constraints generated by the design method is presented.

The expert system aims to provide a flexible design environment, catering for both novice and experienced users. In addition to assisting the designer in using the CACSD package, and unlike previous expert systems for control system design, it also aids the designer in formulating a comprehensive and achievable specification, and in dealing with conflicting design constraints. The expert system has also been useful in effectively expanding the scope of the CACSD package. Chapters 4 and 5 provide details of the expert system concepts and implementation.

Examples illustrating the capabilities of the design system are given in chapter 6. The design system has also been used by students at the University of Cape Town, and chapter 7 contains details of these experiences. Chapter 8 describes two methods for implementing the controllers synthesized by the design system.

Chapter 9 concludes the volume with a review of the most important material discussed.

The research described in this volume was conducted at the Department of Electrical and Electronic Engineering of the University of Cape Town, South Africa. In this regard I would like to express my thanks to Professor Martin Braae for encouragement and constructive criticism during this research, and to thank AECI (Pty) Ltd., and the Foundation for Research Development, for generous financial support. Finally, thanks to Anne Sargent of the University of Detroit library for tracking down numerous conference papers.

Colin Tebbutt
LinkData
PO Box 23316
Claremont 7735
South Africa

TABLE OF CONTENTS

1 Artificial Intelligence and Control System Design 1

1.1 Introduction 1
1.2 Why Use an Expert System? 2
1.3 Applications of Expert Systems in CACSD 3
1.4 Description of the Design System 4
1.5 Notation 7

2 The CACSD Method 9

2.1 Introduction 9
2.2 Notation 10
2.3 Summary of Factorization Theory 10
2.4 Computing the Coprime Matrix Fractions 12
2.5 Diagonal Factorization 13
2.6 The Diagonal Factorization Algorithm 17
2.7 The QSTEP Parameter 18
2.8 The Closed Loop Poles 22
2.9 The S Domain 24
2.10 Summary 24

3 Implementation of the CACSD Package 25

3.1 Introduction 25
3.2 Generating the Quadratic Programming Problem 26
    3.2.1 Computing the Matrix Fractions 26
    3.2.2 Translating the Performance Constraints 27
    3.2.3 Translating the Cost Function 29
3.3 Representing the Linear Constraints 30
    3.3.1 Time Domain Constraints 31
    3.3.2 Frequency Domain Constraints 31
    3.3.3 Examples 34
3.4 Solving the Quadratic Programming Problem 36
3.5 The QPSOL Algorithm 37
3.6 Ancillary Functions of the CACSD Package 39
    3.6.1 Entry of Transfer Function Matrices 39
    3.6.2 Entry of Performance Constraints 39
    3.6.3 Graphics Facilities 39

	3.6.4 Analysis Facilities	40
	3.6.5 Reduced Order Controller Estimation	40
3.7	Summary	40

4 The Expert System — 41

4.1	Introduction	41
4.2	The User Interface and Philosophy	42
4.3	Explaining the Design Language and Methodology	42
4.4	Presentation of the Design Status	44
4.5	Formulating the Design Specifications	44
	4.5.1 The NEXT STEP Command	45
	4.5.2 The SUGGEST Command	46
	4.5.3 The COMPLETE Command	48
4.6	Expanding the Scope of the CACSD Package	49
4.7	Optimizing the Use of the CACSD Subroutines	50
4.8	Summary	50

5 Implementation of the Expert System — 51

5.1	Introduction	51
5.2	Selection of the Expert System Shell	51
5.3	Communication with External Programs	52
5.4	Database Facilities	53
	5.4.1 The RESP–DB Database	53
	5.4.2 The SPEC–DB Database	53
	5.4.3 The SOLVE–DB Database	55
	5.4.4 The PARAM–DB Database	55
	5.4.5 The FEATURE–DB Database	56
5.5	Structure of the Expert System	58
	5.5.1 Initialization	58
	5.5.2 The Command Line Processor	59
	5.5.3 The HELP Module	59
	5.5.4 The SUGGEST and NEXT STEP Modules	59
	5.5.5 The Module Checking for Completeness	63
	5.5.6 The Module Explaining Specification Conflicts	63
5.6	Summary	66

6 Sample Design Sessions — 67

6.1	Introduction	67
6.2	Example 1: A Mine Milling Plant	67
6.3	Example 2: A Gyroscope	80
6.4	Other Examples	83
6.5	Summary	83

7 Use of the Expert System ... 85

7.1 Introduction ... 85
7.2 The Undergraduate Control System Design Project ... 85
7.3 The Postgraduate Control System Design Project ... 87
    7.3.1 Single Variable Designs ... 88
    7.3.2 INA/DNA Designs ... 89
    7.3.3 Characteristic Loci Designs ... 90
    7.3.4 MV-CXS Designs ... 91
    7.3.5 Comparison of Design Methods ... 92
    7.3.6 Alternative Designs ... 95
7.4 Summary ... 96

8 Implementing the Controller ... 97

8.1 Introduction ... 97
8.2 Full Controller Implementation ... 97
8.3 Reduced Order Controller Estimation ... 100
8.4 Examples of Controller Implementation ... 101
    8.4.1 FIR Implementation ... 101
    8.4.2 Reduced Order Controller Estimation ... 103
8.5 Summary ... 105

9 Conclusions ... 97

Appendix A MV-CXS Specifications ... 97

A.1 Requirements for the Computer ... 109
A.2 Requirements for the Plant ... 109
A.3 The MV-CXS Command Language Specification ... 109
    A.3.1 Commands to Select a Response ... 110
    A.3.2 Graphics Commands ... 111
    A.3.5 Miscellaneous Commands ... 115

Appendix B The CACSD Package Interface ... 117

Appendix C The MV-CXS Student Design Project Instructions ... 123

References ... 125

Index ... 131

EDITORIAL BOARD

Professor Dr -Ing J.Ackermann
DLR Institut für Robotik und
Systemdynamik
Postfach 1116
D-82230 Weßling
Germany

Professor I. D. Landau
Le Directeur
Laboratoire d'Automatique de
Grenoble
ENSIEG, BP 46
38402 Saint Martin d'Heres
France

Dr D. C. McFarlane
BHP Research
Melbourne Research Laboratories
245-273 Wellington Road
Mulgrave
Victoria 3170
Australia

Professor B. Wittenmark
Department of Automatic Control
Lund Institute of Technology
PO Box 118
S-221 00 Lund
Sweden

Dr D. W. Clarke, MA., D.Phil,
CEng, FIEE
Reader in Information
Engineering
Department of Engineering
Science
University of Oxford
Parks Road
Oxford, OX1 3PJ
U.K.

Professor H. Kimura
Professor of Control Engineering
Department of Mechanical
Engineering for Computer
Controlled Machinery
Faculty of Engineering
Osaka University
2-1 Yamadaoka
Suita
Osaka 565
Japan

Professor A. J. Laub
Professor and Chairman
Department of Electrical and
Computer Engineering
University of California
Santa Barbara
California 93106
U.S.A.

Professor J. B. Moore
Department of Systems
Engineering
The Australian National
University
Research School of Physical
Sciences
GPO Box 4
Canberra
ACT 2601
Australia

Professor Dr -Ing M. Thoma
Institut Für Regelungstechnik
Universität Hannover
Appelstrasse 11
D-30167 Hanover 1
Germany

Symbols and Abbrebiations

AI	Artificial Intelligence.
C	The set of complex numbers.
C^n	The set of complex vectors of dimension n.
C$_-$	A subset of the complex plane defined as the region of stability; the set $\{z: |z| <1\}$ is often chosen.
C$_{+e}$	The complement of **C**$_-$, including the point at infinity.
CAD	Computer Aided Design.
CACSD	Computer Aided Control System Design.
DNA	Direct Nyquist Array.
F	The field of fractions associated with **S**; this is the set of real rational transfer functions.
FIR	Finite Impulse Response.
I	The unit or identity matrix.
INA	Inverse Nyquist Array.
LCM	Lowest Common Multiple.
LQG	Linear Quadratic Gaussian.
LTR	Loop Transfer Recovery.
M(S)	The set of matrices with elements in **S**, i.e. the set of matrices where the elements are proper stable transfer functions.
M(F)	The set of matrices with elements in **F**, i.e. the set of matrices where the elements are transfer functions.
MIMO	Multi-Input Multi-Output.
R	The set of real numbers.
R^n	The set of real vectors of dimension n.
R[z]	The set of polynomials in the indeterminate z, with real coefficients.
R(z)	The field of fractions associated with **R**[z]. This is the set of real rational transfer functions.
RAM	Random Access Memory.
S	The subset of **R**(z) comprising all rational functions analytic on **C**$_{+e}$. This is the set of proper stable transfer functions.
SISO	Single-Input Single-Output.
SVD	Singular Value Decomposition.
U	The set of units in **S**, i.e. the set of proper stable transfer functions whose inverses are also proper stable transfer functions.
U(S)	The set of unimodular matrices in **M(S)**, i.e. those matrices in **M(S)** whose inverses are also members of **M(S)**

Trademarks

CTRL-C is a trademark of Systems Control Technology.
dmX is a trademark of Decision Management Software.
IBM-PC is a trademark of International Business Machines Corporation.
Personal Consultant is a trademark of Texas Instruments, Inc.
Macsyma is a trademark of Symbolics Inc.
MS-DOS is a trademark of Microsoft Corporation.
Synapse is a trademark of Hitep.
VP-Expert is a trademark of Paperback Software International.

CHAPTER 1
ARTIFICIAL INTELLIGENCE AND CONTROL SYSTEM DESIGN

1.1. Introduction.

The study of artificial intelligence has enjoyed much attention over the past few decades, and the technology has been applied in a wide variety of fields. In control system engineering, artificial intelligence techniques have been employed in two main categories: the online and offline applications. Real-time expert systems [1] and neural networks [2], for example, have been used to implement physical controllers or supervisory control systems. Applications in the second category include system identification, for example [3] and [4], and control system design.

Despite all the attention received, artificial intelligence as a subject still lacks an adequate definition. One popular definition, as expressed by Graham [5], is :

> *"Artificial Intelligence is the branch of computer science devoted*
> *to programming computers to carry out tasks that if carried*
> *out by human beings would require intelligence."*

This definition is hardly satisfactory in the sense that the computation of the first five significant digits of $\sqrt{2}$ requires a great deal of human intelligence, while a simple hand-held calculator, performing this task with breathtaking speed, is rarely considered intelligent. Conversely, it is a simple matter for a human to identify a friend in a crowd, but the same task is formidable even for present day intelligent machines. Clearly humans and computers have differing "natural" abilities and skills.

Trying to account for this, Rich [6] proposes an alternative definition :

> *"Artificial Intelligence is the study of how to make computers do*
> *things at which, at the moment, people are better."*

The design of a control system certainly requires intelligence in a human, and is generally considered to fall within the scope of artificial intelligence, fitting both definitions. Engineering design is a combination of (human) creativity and intelligent decision making [7]. However the aim of any intelligent design system is to produce a first-rate design, and not necessarily to replace the designer. Dreyfus and Dreyfus [8,9] argue persuasively against relying on the skills of the computer alone. The designer and intelligent design system should rather be viewed as a unit or team, where each member contributes specific skills while working towards a common goal.

Pang and MacFarlane [10] list some of the relative strengths and weaknesses of the human and the computer. Humans, for example, have powerful abstraction and pattern-recognition faculties, while computers can perform complex calculations with speed, accuracy and reliability. A combination of human and machine could go a long way towards eliminating the weaknesses of either. Finding this combination should be one of the goals of artificial intelligence research.

The intelligent design system described here, named MV-CXS, is based on this teamwork principle. Its purpose is to assist the user in formulating a comprehensive design specification, to deal with possible conflicts in the design constraints, and to find a controller which meets these specifications. The designer is ultimately responsible for the engineering decisions.

Despite the power of present-day computers, excessive demands on system resources, such as processing time or memory, may render a certain task more suitable for human solution. For example, a specific algorithm may enable a computer to decipher a hand-written letter, but with less speed and/or lower accuracy than a human. In this instance, improving the efficiency or accuracy of the algorithm dramatically would qualify as artificial intelligence work using Rich's [6] definition. In many ways intelligence may be related to speed of response; a person who correctly answers a question quickly is often considered more intelligent than another who takes longer to answer. Similarly, the "intelligence" of two equivalent algorithms could be related by comparing their efficiencies.

A sizable portion of the work reported here relates to improving the time and space efficiency of a quadratic programming algorithm. This algorithm attempts to find the vector **x** which minimizes the quadratic cost function

$$J(\mathbf{x}) = \mathbf{x}^T \mathbf{A} \mathbf{x} + \mathbf{b}^T \mathbf{x} + c,$$

while satisfying many (typically hundreds or thousands of) linear constraints each with the form

$$\mathbf{d}^T \mathbf{x} \geq e.$$

Although this task would certainly require intelligence if performed by a human, the algorithm does not fit Rich's [6] definition, and is not commonly classed as an artificial intelligence technique. Nevertheless its efficiency is vital to usefulness or intelligence of the overall design system, in that for a given computer system it affects the rate at which problems are solved, and determines the upper limit on the size of problems which may be addressed. Similar considerations apply to the "intelligent" (efficient) search strategies, such as the alpha-beta method [6] frequently used in games programs, which have been pervasive in traditional AI research. In both cases knowledge of the specific problem is used to reduce the search space and storage requirements, and accelerate its solution.

1.2. Why Use an Expert System?

There have been two distinct trends in the evolution of control engineering software. Firstly, computer aided design systems have grown from mainly single-purpose programs

Chapter 1. Artificial intelligence and control system design.

for analysis and design [11] into comprehensive packages covering a wide range of control engineering activities [12]. Secondly, the design systems are being aimed at an increasingly wide range of users, and not only at experienced designers [13]. Progress, particularly in the latter direction, has often been based upon expert system technology. While existing computer aided control system design (CACSD) tools are almost exclusively analysis packages [14], the use of expert system techniques have offered the hope of producing fully-fledged design packages, which are able to provide meaningful guidance for the user during the design process.

Expert system techniques [15,16,17,18] deal effectively with the problem of complexity management. In this application, as with many artificial intelligence problems, there is a complex decision making structure and a large amount of knowledge. Expert systems offer a powerful facility for representing decision making knowledge in terms of rules and facts. According to Taylor and Frederick [12],

> "...a rule-based expert system can be endowed with greater flexibility
> than conventional software, because of its knowledge base,
> inference capability, and more natural interaction with the user".

There is no such thing as an instant human expert; in practice every expert goes through a learning phase. Similarly it can be expected that an intelligent design system will go through a similar development cycle. One of the attractive features of the expert system methodology is that, as experience with the system grows, the knowledge base can be extended with relative ease, and this may usually be done without disturbing the structure of the knowledge already present. Again, Taylor and Frederick [12] comment

> "...an expert system is easier to expand than a conventional
> program, in the sense that the mechanics of adding rules that
> embody "new expertise" is straightforward".

Expert system shells usually offer good online debugging aids; for example one may trace the reasoning process to determine how the value was inferred for a particular variable. Effective debugging facilities are a vital component of any complex software project.

1.3. Applications of Expert Systems in CACSD.

Many applications of expert systems in the control system design have been reported to date. A selection of these are discussed below.

Taylor and Frederick [12] (1984), and James et al. [19] (1986), present an overview of the application of expert systems to control engineering, in particular outlining the wide range of design activities which should be addressed. They propose an architecture where the expert system coordinates and integrates many analysis and design procedures.

James et al. [20] (1987), describe an expert system implementation of an algorithm for single variable lead-lag compensator design. This in turn forms a small part of Taylor and Frederick's [12] expert system mentioned above, and interfaces to subroutines in the Cambridge Linear Analysis and Design Program CLADP [21]. After specifying the performance required, the designer has little further involvement in the design process.

Trankle *et al.* [22] (1986) describe a two-level expert planning system, based upon the CTRL-C package [23]. A high level planner produces a skeleton plan, and a list of performance specifications for the low level planner to satisfy. The low level planner in turn creates a list of commands for the CTRL-C package. Side effects, resulting from the tradeoffs inherent in controller design, complicate the planning. Again, the user has minimal involvement in the design process.

Nolan [24] (1986) developed an expert system which deduces the feasibility of various single variable feedback configurations, as well as an appropriate synthesis method to be used, based upon the plant type number and order. The system uses algebraic manipulation of the plant transfer function and controller structure, and once a synthesis method is selected, guides the designer in using an appropriate conventional CAD package.

Birdwell *et al.* [25] (1985) discuss an expert system interface to a multivariable LQG/LTR design package. The expert system, named CASCADE, was found useful in automating some of the design details, thus allowing the user to concentrate on the design process and significantly reducing the time required to complete a trial design [26].

The MAID expert system described by Pang *et al.* [10,27] (1987), and Boyle *et al.* [28] (1989), uses three design techniques to address the design problem. These are their "simple design technique", a "reverse frame alignment technique", and an observer-based controller design technique, in order of increasing complexity. Data for design analysis is obtained from the characteristic gains and phases and principal gains of the system. The design is initially attempted using the simpler methods, and if not satisfactory, the more complex methods are employed. The authors emphasize interactive design, and recommend that engineering judgment decisions be left to the designer. This design system has recently been extended to include a stable factorization design method not unlike that used here (Pang *et al.* [14], 1990).

1.4. Description of the Design System.

Each of the expert system based design systems mentioned above attempts to find a controller which satisfies some design specification. The expert system described here takes this approach one step further by also assisting the designer in developing the specification. As stressed by MacFarlane *et al.* [29], design is an exploratory and experimental process during which the specification is systematically refined. Therefore it is appropriate that an intelligent design tool should address the evolution of the specification during this process.

The overall structure of the MV-CXS design system illustrated in figure 1.1 shows an expert system interposed between the designer and a CACSD package; an essentially similar structure is found in each of the applications mentioned above. In broad functional terms, the expert system assists the designer to formulate the design specification, and the CACSD package attempts to synthesize a controller which meets those specifications. Some commands from the designer, for example those for plotting the control system responses, pass directly through the expert system to the CACSD package (figure 1.2(a)), in line with the "command spy" concept of Larsson and Persson [4], and functionally equivalent to the structure used by Pang and MacFarlane [10] (figure 1.2(b)). Other commands, for example the SOLVE command, are translated into a sequence of calls to the CACSD package. Still others, for example the HELP commands, do not (at least directly) result in any calls to the CACSD package.

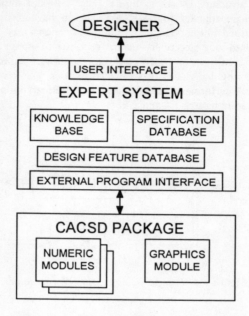

Figure 1.1. Structure of the design system.

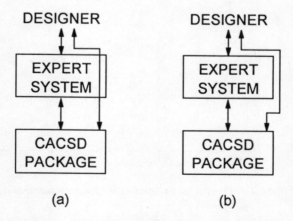

Figure 1.2. Possible relationships between the designer, expert system and CACSD package.

The CACSD package is based upon the design method of Boyd *et al.* [30,31], which is in turn similar to that of Fegley [32], and in theory to that of Gustafson and Desoer [33,34]. It is applicable to linear time-invariant multivariable sampled data systems using the two-parameter controller structure shown in figure 1.3. The plant must be described by a strictly proper z-domain transfer function, and should have a diagonal, left-coprime factorization as described in chapter 2. Design specifications may include performance constraints on the closed loop step or frequency responses to inputs at nodes R, N, or D, as well as a quadratic cost function based on these responses. Further specifications, such as constraints on the singular values of various responses, are treated implicitly by the expert system. Typical performance constraints on the closed loop step and frequency responses are illustrated in figures 1.4 and 1.5.

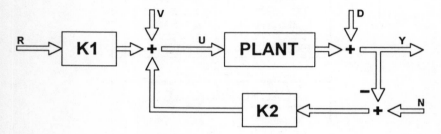

Figure 1.3. Two-parameter controller structure.

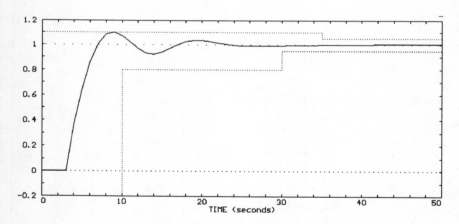

Figure 1.4. Typical time domain performance constraints.

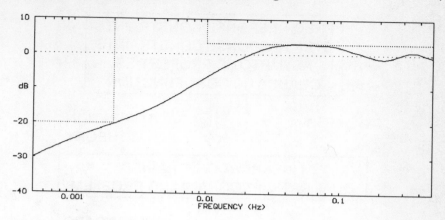

Figure 1.5. Typical frequency domain performance constraints.

The CACSD design method transforms the control system design problem into a linearly constrained quadratic programming problem, which is then solved using a standard algorithm. The solution, if any exists, is then transformed into a controller transfer function using an explicit formula. Figure 1.6 illustrates the important steps in this design process. Unlike most other design methods for multivariable systems, the complexity of the design procedure, as seen by the designer, does not increase dramatically as the dimension of the plant increases. The designer does have more trade-off options to consider, but the bulk of the additional complexity is absorbed by the CACSD package. Many conceptually similar constrained optimization methods for control system design have been proposed elsewhere, for example [35,36,37]. The advantage of Boyd's method [30], however, is that the quadratic programming algorithm always finds the global optimum solution where one exists; the absence of any feasible solution is also determined in a finite number of iterations. This advantage is a result of convexity of the problem; the optimization function, being quadratic, has no local minima which are not also the global minimum, and only performance constraints which can be translated into a convex set of linear constraints on the search vector are allowed. Linear programming, closely related to quadratic programming, has also found applications in control system design, and formed the basis of Fegley's work [32]. More recently it has also been used by Bhattacharyya *et al.* [38,39].

1.5. Notation.

In mathematical descriptions, matrices are shown in bold type with upper case names, for example **A**; vectors are also shown in bold type, but with lower case names, for example **b**. **R** is used to denote the set of real numbers, and **C** the set of complex numbers. Some additional mathematical notation is required in chapter 2, and is introduced there.

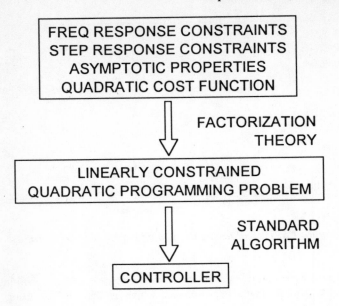

Figure 1.6. Steps in the CACSD design method.

The plant is assumed to have n outputs and n inputs, and is represented by the transfer function matrix $\mathbf{G}(z)$. Thus all other transfer function matrices will have the same dimensions. The theory underlying this design system extends readily to plants which are not square. In many cases, the dependence of transfer functions on the variable z is not shown explicitly; for example the plant is often represented simply as $\mathbf{G}$.

During its interaction with the designer, the expert system and CACSD package refer to the various closed loop responses of figure 1.3 using a two letter notation; for example, DY indicates the closed loop response to stimuli at input D, as observed at output Y. Individual elements in a response matrix are identified using the notation [i,j]; thus RU[2,3] denotes the response at output U2 to stimuli at input R3. In addition, the plant transfer function matrix is denoted by $\mathbf{G}$, those of the controller by $\mathbf{K1}$ and $\mathbf{K2}$, and the open loop response (i.e. $\mathbf{GK2}$) by $\mathbf{GK}$. A similar notation is used here. The closed loop transfer functions are referred to as $\mathbf{H}_c$, where c indicates the input and output nodes. For example, $\mathbf{H}_{RY}$ is the closed loop transfer function from R to Y. The corresponding time domain response at the output node to a unit step at the input node is indicated by $\mathbf{h}_c(kT)$, where T is the sampling time.

Frequency domain responses are evaluated at a number of points logarithmically spaced on the unit circle in the complex plane. Graphical plots of these responses are displayed using linear interpolation between successive frequency points. Similarly, plots of time domain responses also use linear interpolation between successive sampling instants. In some instances the plots of the multivariable responses have been superimposed.

CHAPTER 2
THE CACSD METHOD

2.1. Introduction.

The CACSD design method of Boyd *et al.* [30] is based on translating the control system design problem into an approximately equivalent linearly constrained quadratic programming problem, finding the solution to this using a standard algorithm, and then translating the solution back to give the corresponding controller. The design is specified directly in terms of performance constraints on the closed loop time and frequency domain responses, and a cost function to be minimized.

This design algorithm turns out to be computationally demanding in terms of both memory size and processing speed, especially when used for multivariable systems. However there are a number of techniques which, while retaining most of the strengths of the original algorithm, ease the computational burden substantially, and make implementation on a low cost personal computer practical. These techniques are described below[1].

After introducing some notation and presenting an overview of factorization theory, this chapter examines methods for computing the stable coprime factorizations of the plant transfer function matrix. For the case where the plant can be stabilized using a stable controller, explicit formulae for these factorizations have been developed. Next it is shown that under certain conditions the multivariable design problem may be divided into a number of smaller independent sub-problems, which greatly reduces the parameter vector dimension. This dimension impacts on the memory needed to store the linear constraints, and on the internal storage requirements of the quadratic programming algorithm, effectively limiting the size of problems which may be addressed. The technique requires a diagonal left-coprime factorization of the plant transfer matrix, and an algorithm to compute this is presented. A second opportunity for reducing the dimension of the parameter vector is found in the parameterization of the search space; this aspect is addressed in the section on the QSTEP parameter. The chapter concludes with a discussion on the closed loop poles and implications of applying these techniques in the s domain.

While some of the techniques described here are not applicable in every design situation, the range of designs which may be tackled remains large.

---

[1]Parts of this chapter and the next (c) IEEE, and are reprinted, with permission, from IEEE Transactions on Automatic Control, vol. 37, no. 10, pp. 1631-1634, 1992.

2.2. Notation.

The notation used below follows that of Vidyasagar [40] closely. Let **R**[z] denote the set of polynomials in the indeterminate z, with real coefficients, and **R**(z) the field of fractions associated with **R**[z]. Define a subset of the complex plane **C**_ as a region of stability; the set $\{z : |z| < 1\}$ is often chosen, and will be assumed for the examples given later. Let **C**$_{+e}$ denote the complement of this region, including the point at infinity. Let **S** denote the subset of **R**(z) comprising all rational functions analytic on **C**$_{+e}$, i.e. the set of proper stable transfer functions. **S** is then a commutative ring with identity, and is a domain. Let **F** be the field of fractions associated with **S**, which is also **R**(z).

Let **M(S)** denote the set of matrices with elements in **S**, and **M(F)** the set of matrices with elements in **F**. Let **U** denote the set of units in **S**, and **U(S)** the set of unimodular matrices in **M(S)**.

In transfer function terminology, **M(F)** is the set of matrices with transfer functions as elements, and **M(S)** is the set of matrices where the elements are proper stable transfer functions. **U** is the set of proper stable transfer functions whose inverses are also proper stable transfer functions, and **U(S)** comprises those matrices in **M(S)** whose inverses are also members of **M(S)**.

This application assumes that the plant to be controlled is a linear time-invariant sampled-data process which is described by a strictly proper z domain matrix transfer function $G(z) \in $ **M(F)**, with n inputs and n outputs. The results presented in this chapter apply equally to non-square plants, and those which are proper (but not necessarily strictly proper).

The closed loop transfer functions are referred to as H_c, where c indicates the input and output nodes shown in figure 2.1. For example, H_{RY} is the closed loop transfer function from R to Y. The corresponding time domain response at the output node to a unit step at the input node is indicated by $h_c(kT)$, where T is the sample time.

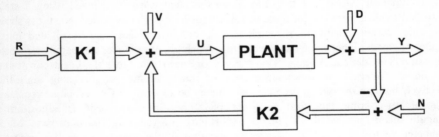

Figure 2.1. Two-parameter control system configuration.

2.3. Summary of Factorization Theory.

A brief summary of the factorization approach to the design of linear control systems is given below; for a thorough treatment of the subject see Vidyasagar [40]. Matrix fraction descriptions were proposed by Rosenbrock [41], and used by Youla *et al.* [42] to parameterize the family of stabilizing controllers. This application is based upon the two-

Chapter 2. The CACSD Method.

degree-of-freedom multivariable control system structure, shown in block diagram form in figure 2.1.

Given stable right- and left-coprime factorizations $N, D \in M(S)$ and $\tilde{N}, \tilde{D} \in M(S)$ of the plant

$$G = ND^{-1} = \tilde{D}^{-1}\tilde{N} \qquad (2.1)$$

there exist matrices $X, Y \in M(S)$ which satisfy the Bezout Identity

$$XN + YD = I \qquad (2.2)$$

The important result of factorization theory follows : *all* internally stable closed loop transfer functions $H_c \in M(S)$ may be parameterized in terms of some $Q \in M(S)$ as

$$H_c = H0_c + H1_c Q H2_c \qquad (2.3)$$

The transfer function matrices $H0_c$, $H1_c$ and $H2_c \in M(S)$ are defined in terms of $N, D, \tilde{N}, \tilde{D}, X$, and $Y \in M(S)$; table 2.1 lists the formulae for each of the closed loop transfer functions. Q is one of two independent parameter matrices Q_1 and Q_2; these two matrices are chosen by the designer to give the required closed loop performance.

Table 2.1. Closed loop transfer functions.

Response (H_c)	$H0_c$	$H1_c$	Q	$H2_c$
H_{RY}	0	N	Q_1	I
H_{RU}	0	D	Q_1	I
H_{NY}	NX	N	Q_2	$\tilde{D}$
H_{NU}	DX	D	Q_2	$\tilde{D}$
H_{DY}	I - NX	-N	Q_2	$\tilde{D}$
H_{DU}	-DX	-D	Q_2	$\tilde{D}$
H_{VY}	NY	N	Q_2	$\tilde{N}$
H_{VU}	DY	D	Q_2	$\tilde{N}$

Having selected values for Q_1 and Q_2 which give the required closed loop performance, the final step of the design process is the computation of the controller from the formulae

$$K_1 = (Y - Q_2\tilde{N})^{-1}(Q_1)$$

and

$$K_2 = (Y - Q_2\tilde{N})^{-1}(X + Q_2\tilde{D}) \qquad (2.4)$$

The matrix

$$K_0 = Y^{-1}X \qquad (2.5)$$

may be thought of as an initial stabilizing controller for the one-degree-of-freedom system shown in figure 2.2. X and Y are then a left-coprime factorization of K_0.

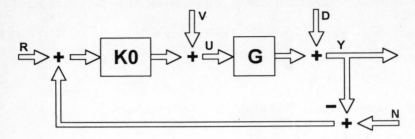

Figure 2.2. One-parameter control system configuration.

The essence of the design method is the computation of the transfer function matrices $H0_c$, $H1_c$ and $H2_c$, and the selection of suitable parameter matrices $Q1$ and $Q2$ to give the required closed loop performance. The latter step is achieved using a parameterization of Q, and the subsequent translation of the design specification into a set of linear constraints on, and a cost function in terms of, these parameters. The quadratic programming algorithm is then used to solve for the parameters, which in turn gives $Q1$ and $Q2$.

2.4. Computing the Coprime Matrix Fractions.

As described above, matrix fractions play a central role in the factorization theory approach. This section examines methods for computing the fractions. For the general case, when a state-space representation of the plant is available together with stabilizing state feedback matrices, formulae for stable coprime fractions of G and $K0$ are available [43]. Zhao and Kimura [44] also give formulae based on the Smith-McMillan form of the plant transfer function matrix. However neither of these sets of equations are necessarily the most convenient to use; in particular there are two special cases which lend themselves to simplified computation of the coprime factorizations.

    1.   **Stable plant.** [45,33]

In this case the factorization is trivial, and no stabilizing controller is required.

$$N = \tilde{N} = G$$
$$D = \tilde{D} = Y = I$$
$$X = 0 \tag{2.6}$$

    2.   **Stable initial controller.**

Many unstable plants can be stabilized using a stable controller $K0 \in M(S)$; Vidyasagar [40] gives the conditions under which this is possible (corollary 5.3.2). In this case a right-coprime factorization $N, D \in M(S)$ can be chosen as

$$N = G(I + K0G)^{-1}$$
$$D = (I + K0G)^{-1}$$
$$X = K0$$
$$Y = I \tag{2.7}$$

Proof

$$ND^{-1} = G(I + K_0G)^{-1}(I + K_0G)$$
$$= G$$

$$XN + YD = K_0G(I + K_0G)^{-1} + (I + K_0G)^{-1}$$
$$= (K_0G + I)(I + K_0G)^{-1}$$
$$= I$$

As K_0 is a stabilizing controller, N and D are stable, representing the closed loop transfer functions from V to Y and V to U in figure 2.2 respectively. Since the Bezout Identity is also satisfied, it follows that N and D are right-coprime ([40] corollary 4.1.17).

□

Similar expressions exist for stable left-coprime $\tilde{N}$ and $\tilde{D}$, and are derived in an analogous fashion; they are :

$$\tilde{N} = (I + GK_0)^{-1}G$$
$$\tilde{D} = (I + GK_0)^{-1} \tag{2.8}$$

While the expressions for N, D, $\tilde{N}$ and $\tilde{D}$ may be quite complex, their algebraic form need not be computed explicitly. In the frequency domain these expressions are easily evaluated at discrete frequencies in terms of the transfer functions G and K_0. In the time domain their impulse responses are required; these may be obtained by simulating the corresponding closed loop systems.

Direct factorization techniques may also be used to compute the matrix fractions, particularly for plants not fitting either of the categories described above. One such technique is the diagonal factorization method described later, which requires only slight modifications to produce a right factorization.

Gustafson and Desoer [34] in fact propose that the frequency and impulse responses of the matrix fractions be obtained directly from plant measurements, without explicitly identifying a model for the plant. For example, the Fourier transform of the impulse response could be used to give the frequency response. Unfortunately, for true multivariable systems, this approach is compatible with the diagonal factorization discussed below only when the plant is stable.

2.5. Diagonal Factorization.

It is interesting to note that if the matrix $H2_c$ is diagonal, then the individual elements of a specific closed loop transfer function can be written as

$$\mathbf{H}_c[i,j] = \mathbf{H0}_c[i,j] + \sum_{k=1}^{n} \mathbf{H1}_c[i,k]\mathbf{Q}[k,j]\mathbf{H2}_c[j,j] \tag{2.9}$$

Thus column j of $\mathbf{H}_c$ depends only on column j of $\mathbf{Q}$, and the design may be reduced from a single design problem of size n^2 to n independent sub-problems each of size n. Size here refers to the dimension of the search vector used by the quadratic programming algorithm. This reduction, when possible, greatly extends the scale of design problems which may be tackled given limited computer memory.

Consider a plant with 4 inputs and 4 outputs, and where each element of $\mathbf{Q}$ has 5 parameters (decision variables). Here the reduction results in 4 sub-problems each of size 20, instead of a single problem of size 80. Each of the linear constraint vectors will require n times as much storage for the single problem, and there will usually be about n times as many of them, increasing the storage requirements by a factor n^2. For example, assume 1000 linear constraints are generated per sub-problem, and 8 bytes are needed per floating point number; then $1000 \times 20 \times 8 = 160$ kilobytes are required to store the constraint vectors for each sub-problem, as opposed to $4 \times 1000 \times 80 \times 8 = 2560$ kilobytes for the single problem. Note that since the sub-problems are solved independently, the constraints for each need not be stored simultaneously. Furthermore the storage requirement for the six matrices in the quadratic programming algorithm is only $20^2 \times 8 \times 6 = 19200$ bytes for each sub-problem, instead of $80^2 \times 8 \times 6 = 307200$ bytes.

The time required to produce the final solution is usually also reduced for the partitioned problem as, although there are n problems to solve instead of just one, each one is very much simpler. An additional advantage of subdividing the problem is that conflicts within the engineering specifications are generally easier to identify and resolve, as there are fewer specifications in each sub-problem.

Unfortunately the partitioning scheme hinges on a diagonal $\mathbf{H2}_c$. The circumstances under which this may be arranged are investigated next.

From table 2.1,

$$\mathbf{H2}_c = \begin{cases} \mathbf{I}, & c \in \{RY, RU\} \\ \tilde{\mathbf{D}}, & c \in \{DY, DU, NY, NU\} \\ \tilde{\mathbf{N}}, & c \in \{VY, VU\} \end{cases}$$

The matrix $\mathbf{I}$ is diagonal by definition. Clearly both $\tilde{\mathbf{N}}$ and $\tilde{\mathbf{D}}$ cannot be diagonal simultaneously, except when the plant is diagonal; this case will not be considered further as the corresponding control problem may be solved using standard single variable methods. Assuming that $\tilde{\mathbf{D}}$ may be chosen to be diagonal, it is then necessary to forgo the opportunity of explicitly designing the closed loop transfer functions $\mathbf{H}_{VY}$ and $\mathbf{H}_{VU}$. This is considered a small sacrifice compared to the advantages of the resulting independence. By comparison, many other multivariable design methods such as the INA and characteristic loci methods allow disturbances to be considered explicitly at either the input or the output of the plant, but not at both simultaneously.

The statements above should not be taken to imply that the designer has no control over the closed loop responses H_{VY} and H_{VU}, only that it will not be possible to design them *explicitly*. Since

$$H_{VY} = H_{DY} G$$

and $\quad H_{VU} = H_{DU} G$

these responses may be designed indirectly, particularly in the frequency domain, by careful shaping of the H_{DY} and H_{DU} responses, bearing in mind the characteristics of the plant which can be thought of as a pre-filter.

Definition 2.1
A matrix factorization will be termed diagonal when the denominator matrix is diagonal.
□

A technique for computing a diagonal factorization of the plant is presented later in the form of a simple algorithm. While the factorization is not always coprime, a simple test is available to determine if the factorization is coprime.

According to Gustafson and Desoer [34] (1985),

> "There is no reliable software available that will perform coprime factorizations, multiplications or additions of matrices over the ring of polynomials or rational functions. Much of the available software suffers from numerical sensitivity and ill-conditioning."

It is expected that the simplicity of the proposed diagonal factorization method will make it less susceptible to these numerical problems. The method has indeed performed well in the applications tested.

The two theorems that follow are based on theorems found in Vidyasagar [40]. While he generally treats only the right-coprime case explicitly, the corresponding left-coprime forms used below follow readily. Theorem 2.1 is derived from problem 4.1.11 of [40].

Theorem 2.1
Let $G \in M(F)$ have a left-coprime factorization $A, B \in M(S)$. Then $G + H$ has a left-coprime factorization $A + BH, B$ for all $H \in M(S)$.

Proof
As A and B are left-coprime, there exist $X, Y \in M(S)$ such that

$$AX + BY = I \qquad (2.10)$$

([40] corollary 4.1.17). Define $Y' \in M(S)$ as

$$Y' = Y - HX. \qquad (2.11)$$

Then

$$(A + BH)X + BY' = AX + BY = I. \qquad (2.12)$$

Thus $(A + BH), B \in M(S)$ are left-coprime ([40] corollary 4.1.17). Furthermore

$$B^{-1}(A + BH) = G + H. \qquad (2.13)$$

□

Theorem 2.2
Let $A, B \in M(S)$ each have n rows, and let the sum of the number of columns of each be at least n. Then A and B are left-coprime if $\text{rank}([A\ B]) = n$ at all points in C_{+e}.

Proof
There exists $U \in U(S)$ such that

$$[A\ B]U = [R\ 0], \qquad (2.14)$$

where $R \in M(S)$ is a greatest common left divisor of A and B ([40] corollary B.2.15). Further there exist $X, Y \in M(S)$ such that

$$AX + BY = R \qquad (2.15)$$

([40] theorem 4.1.7).

Since U is unimodular, $\det(U) \in U$ ([40] fact B.1.26), and thus $\det(U)$ is nonzero at all points in C_{+e}. Thus U is nonsingular in C_{+e}, and

$$\text{rank}([A\ B]) = \text{rank}(R). \qquad (2.16)$$

If $\text{rank}([A\ B]) = n$ in C_{+e}, then $\det(R)$ is nonzero in C_{+e}. Note that $\det(R) \in S$ by definition of the determinant, and therefore $\det(R)$ has neither poles nor zeros in C_{+e}. Consequently $\det(R) \in U$, $R \in U(S)$ ([40] fact B.1.26), and $R^{-1} \in M(S)$.

Multiplying (2.15) on the right by R^{-1} gives

$$AXR^{-1} + BYR^{-1} = RR^{-1} = I. \qquad (2.17)$$

As $XR^{-1}, YR^{-1} \in M(S)$, it follows that A and B are left-coprime ([40] corollary 4.1.17).

Remarks
Although not required here, an "only if" clause for this theorem can also be proved. A similar theorem, using $R[z]$ in place of S, is found in Kailath [46].

□

2.6. The Diagonal Factorization Algorithm.

The two theorems in the preceding section form the basis of the diagonal factorization algorithm. The first step in this procedure is to split the plant into stable and unstable components

$$G = G_s + G_u \tag{2.18}$$

such that $G_s \in M(S)$, and $G_u \in M(F)$ contains only the unstable poles of G. Then, according to theorem 2.1, if G_u has a diagonal left-coprime factorization, so does G.

A left factorization of G_u

$$G_u = \tilde{D}^{-1}\tilde{N}_u \tag{2.19}$$

with $\tilde{N}_u, \tilde{D} \in M(S)$ and $\tilde{D}$ diagonal, may be produced as follows:

let $\quad G_u[i,j] = a_{ij} / b_{ij} \quad\quad\quad i, j \in \{1, 2, ...n\} \tag{2.20}$

and $\quad d_i = \text{LCM } b_{ik}, \quad\quad\quad i, k \in \{1, 2, ...n\} \tag{2.21}$

where $a_{ij}, b_{ij}, d_i \in R[z]$. Choose some polynomial $c_i \in R[z]$ with the same order as d_i, and all of its zeros in C_-; then define the elements of the denominator matrix as

$$\tilde{D}[i,i] = d_i / c_i \quad\quad\quad i \in \{1, 2, ...n\} \tag{2.22}$$

Finally the numerator matrix is given by,

$$\tilde{N}_u = \tilde{D} G_u$$

and $\quad \tilde{N} = \tilde{D} G \tag{2.23}$

Finally it is necessary to employ theorem 2.2 to determine if this factorization is left-coprime. In this case the application of the theorem is eased by the diagonal structure of $\tilde{D}$. $[\tilde{N}_u \ \tilde{D}]$ clearly has full rank in C_{+e}, except (possibly) at the unstable poles of G (the zeros of $\tilde{D}$). At each of these poles z_u, full rank is possible if the rows of $\tilde{N}_u(z_u)$ corresponding to those of $\tilde{D}(z_u)$ which are now zero, are linearly independent.

As an example, consider the unstable plant

$$G(z) = \begin{bmatrix} \dfrac{-10.517}{z-1.1052} & \dfrac{-10.517}{z-1.1052} \\ \dfrac{-10.517}{z-1.1052} & \dfrac{-11.070}{z-1.2214} \end{bmatrix} z^{-2}$$

examined in [44]. Applying the technique above gives the diagonal left factorization

$$\tilde{D}(z) = \begin{bmatrix} \dfrac{z-1.1052}{z-0.9} & 0 \\ 0 & \dfrac{(z-1.1052)(z-1.2214)}{(z-0.9)^2} \end{bmatrix}$$

$$\tilde{N}(z) = \begin{bmatrix} \dfrac{-10.517}{z-0.9} & \dfrac{-10.517}{z-0.9} \\ \dfrac{-10.517}{(z-0.9)^2} & \dfrac{-11.070(z-1.1052)}{(z-0.9)^2} \end{bmatrix} z^{-2}$$

Evaluating $[\tilde{N} \ \tilde{D}]$ at the unstable poles $z = 1.1052$ and $z = 1.2214$ respectively gives

$$\begin{bmatrix} \tilde{N} & \tilde{D} \end{bmatrix} = \begin{bmatrix} -41.96 & -41.96 & 0 & 0 \\ 23.76 & 0 & 0 & 0 \end{bmatrix}$$

and

$$\begin{bmatrix} \tilde{N} & \tilde{D} \end{bmatrix} = \begin{bmatrix} -21.93 & -21.93 & 0.36 & 0 \\ 0 & -8.34 & 0 & 0 \end{bmatrix}$$

respectively. Both of these matrices have full row rank, and thus $\tilde{N}$ and $\tilde{D}$ are left-coprime.

2.7. The QSTEP Parameter.

A result from factorization theory is that all stable closed loop transfer functions may generated by equation 2.3 for some stable transfer function matrix $Q \in M(S)$. When the domain of Q is restricted, this fact no longer holds. Unfortunately some restriction is inevitable when Q is represented on a finite computer, and is potentially very serious when Q is parameterized using a small number of decision variables.

To produce an acceptable engineering solution it is generally not essential that all possible transfer functions be generated; a representative range suffices. Consider a typical

Chapter 2. The CACSD Method.

computer representation of the set of real numbers, where both the range of numbers, as well as the precision of the representation, is limited; yet for most problems this set of values available is perfectly adequate. Similarly it is required that the set of possible values for **Q** spans an adequate subset of all stable transfer function matrices, and with adequate precision.

The design method of Boyd [30] is based upon a finite impulse response (FIR) representation of **Q**; in essence the method requires a matrix of proper stable polynomial ratios, where the denominator polynomials are fixed, and the coefficients of the numerator polynomials are determined by the search algorithm. To what extent does this representation of **Q** approximate the set of all stable transfer functions? For low order filters, it would seem rather poorly.

A clue to the physical meaning of **Q** is obtained from the case where the plant is stable. Choosing the factorization of equation 2.6 gives

$$\mathbf{H}_{RU} = \mathbf{Q}_1 \tag{2.24}$$

and $\quad \mathbf{H}_{NU} = \mathbf{Q}_2 \tag{2.25}$

Here **Q**$_1$ and **Q**$_2$ are required to represent the closed loop transfer functions to the plant input. In the time domain it is clearly seen that the order of the FIR filter marks a time window over which control actions, following a disturbance impulse, may be taken. A high order filter is therefore necessary if a "slow" (in terms of the sampling rate) control is desired. Unfortunately the dimension of the search vector in the quadratic programming algorithm is directly proportional to the order of the FIR filter, which makes the use of high order filters prohibitive on a small computer. In order to gain the computational advantages of using a low dimension search vector, and yet retain some of the benefits of a high order FIR filter, the following (first order) parameterization was proposed [47]:

$$\mathbf{Q}[i,j](z) = q_{0,ij} + \sum_{k=1}^{p-1} q_{k,ij} \sum_{r=1}^{QSTEP} z^{-(r+(k-1)QSTEP)} \tag{2.26}$$

where $q_{k,ij}$, $k \in \{0, 1, \ldots p\text{-}1\}$, are the coefficients (decision variables) of the new filter **Q**[i,j]. The number of decision variables per element of **Q**, p, is also referred to as NVARS by the expert system. QSTEP is a positive integer which effectively stretches the FIR filter; the standard FIR form results when QSTEP is unity. Generally QSTEP is chosen in the light of the required speed of response (relative to the sampling rate).

This first order parameterization performs well in many cases; however the control signal from designs of this form often exhibit undesirable high frequency properties, seen as sharp changes in the control action. To resolve this problem, a second order parameterization has been developed, which gives a much smoother control action.

As before, let $q_{k,ij} \in \mathbf{R}$, $k \in \{0, 1, \ldots p\text{-}1\}$, be the coefficients of the filter **Q**[i,j]. Then the filter can be parameterized as

$$Q[i,j](z) = q_{0,ij} + \sum_{k=1}^{p-1} q_{k,ij} \sum_{r=1}^{QSTEP} \alpha_{k,r} z^{-(r+(k-1)QSTEP)} \qquad (2.27)$$

where

$$\alpha_{k,r} = \begin{cases} (QSTEP)q_{k,ij} & \text{when } k = 1 \\ (QSTEP-r)q_{k-1,ij} & \text{when } k = p \\ (r)q_{k,ij} + (QSTEP-r)q_{k-1,ij} & \text{elsewhere} \end{cases}$$

Figure 2.3 shows the effective weighting functions for the first and second order parameterizations. These are given in the form of FIR filters qp[k], $k \in \{0, ... p\}$, for each of the parameters $q_{k,ij}$, for the case where p = 4, and QSTEP = 3. Thus the composite FIR filter is given by

$$Q[i,j](z) = \sum_{k=1}^{p-1} q_{k,ij} \sum_{t=0}^{\infty} qp[k](t) z^{-t} \qquad (2.28)$$

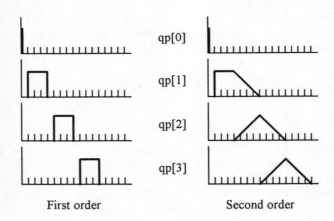

Figure 2.3. Weighting functions for the first and second order Q parameterizations.

The merit of the QSTEP parameter is most clearly illustrated by a single-input single-output example. Consider the plant, sampled at a rate of 1 Hz, with transfer function

$$G(z) = \frac{0.01}{(z-0.8)(z-0.95)}$$

Chapter 2. The CACSD Method.

Let the design problem be the minimization of the cost function J based on the response to a unit step disturbance at the plant output,

$$J = \sum_{k=0}^{50} h_{DY}(kT)^2$$

subject to the constraint on the control signal

$$|h_{DU}(kT)| \leq 5.0, \qquad k = 0, 1, \ldots 50$$

and the asymptotic disturbance rejection requirement

$$H_{DU}(1) = 0.$$

Table 2.2 lists the minimum value found for the cost function J for different combinations of the number of decision variables p and the value of QSTEP. Using 5 decision variables, the minimum cost is obtained with QSTEP set at 4, and this cost is only slightly higher than that of the best long FIR filter. Figure 2.4 shows the step responses for the a short FIR filter (response a: p=5, QSTEP=1), a short FIR filter stretched by the QSTEP parameter (response b: p=5, QSTEP=4), and a long FIR filter (response c: p=25, QSTEP=1). Note that linear interpolation is used between the output values at the sampling instants in this and the other graphs.

Table 2.2. Design cost for various values of p and QSTEP.

p	QSTEP	Cost
5	1	5.442
5	2	5.008
5	3	4.960
5	4	4.936
5	5	5.140
10	1	4.915
15	1	4.910
25	1	4.910

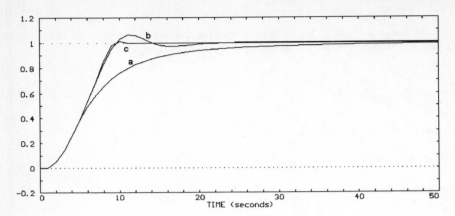

Figure 2.4. Step responses for various values of p and QSTEP.

This SISO example is also used to illustrate the difference between the first and second order parameterizations. Using the first order approximation with p set at 5, the lowest cost (4.948) is also achieved with QSTEP = 4. Figure 2.5 shows the control signals generated using the first (a) and second (b) order parameterizations, with p = 5 and QSTEP = 4. It can be seen that the second order parameterization gives a much smoother control action.

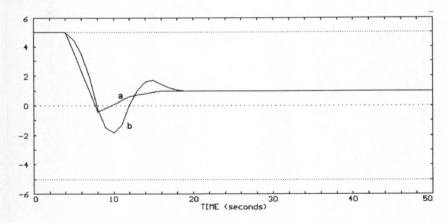

Figure 2.5. Control signal for the first (a) and second (b) order Q parameterizations.

2.8. The Closed Loop Poles.

From equation 2.3 and table 2.1 it is clear that the closed loop poles are given by those of the stable transfer function matrices $\mathbf{N}$, $\mathbf{D}$, $\mathbf{\tilde{N}}$, $\mathbf{\tilde{D}}$, $\mathbf{X}$, $\mathbf{Y}$, and $\mathbf{Q}$. Nevertheless there is a

Chapter 2. The CACSD Method.

considerable amount of freedom available for choosing the closed loop pole positions, which in turn results from the choice of the nominal stabilizing controller **K0**, the transfer function matrix **Q**, and the left factorization **Ñ** and **D̃**.

The choice of **K0** (equation 2.5) determines the poles for **H0**$_c$ (the nominal closed loop system) in equation 2.3. When equation 2.7 is used to compute the matrix fractions **N** and **D**, the choice of **K0** also determines the poles for **H1**$_c$. With stable plants it is common to use equation 2.6 to compute **N** and **D** (**K0** = 0); in this case the poles of the plant will be amongst the closed loop poles.

The diagonal factorization described in section 2.5 also offers some freedom for choosing the poles. The zeros of the polynomial c_i, which become poles of **D̃** (equation 2.22) and thus of **H2**$_c$ and the closed loop system, may be chosen freely. The CACSD package computes the polynomial c_i based on the zeros of d_i (equation 2.21) and a maximum pole modulus parameter β specified by the designer, with $0 < \beta < 1$. Let d_i have the form

$$d_i = a_i \prod_k (z - p_{i,k}) \tag{2.29}$$

with $\quad p_{i,k} = r_{i,k} e^{\theta_{i,k}} \tag{2.30}$

Since d_i contains only the unstable poles of the plant, $r_{i,k} \geq 1$. Then c_i is chosen by the CACSD package as

$$c_i = \prod_k (z - s_{i,k}) \tag{2.31}$$

where

$$s_{i,k} = \begin{cases} (r_{i,k})^{-1} e^{\theta_{i,k}} & (r_{i,k})^{-1} \leq \beta \\ \beta e^{\theta_{i,k}} & (r_{i,k})^{-1} > \beta \end{cases} \tag{2.32}$$

The choices mentioned above are virtually irrelevant when **Q** can span the entire set of stable transfer function matrices, since the stable poles of **H1**$_c$ and **H2**$_c$ can be canceled by zeros of **Q**. Unfortunately this is not generally true when using any finite computer representation for **Q**. Here the poles of **Q**, which may be chosen by the designer, appear as poles of the closed loop system, and the poles of **H1**$_c$ and **H2**$_c$, and those of **Q** itself, cannot always be canceled by zeros of **Q** (which are selected by the design algorithm). This application, as well as that described by Boyd [30], use a FIR filter parameterization for **Q**, which has poles at $z = 0$ only.

This entire section should be tempered by the knowledge that the importance of the pole positions diminishes as the number of poles increases, and systems designed using this method have a large number of poles. A 50 tap FIR filter provides an effective illustration of this principle; an extremely wide range of responses may be generated by varying the filter coefficients, while the pole positions remain fixed at $z = 0$. Boyd [30] gives a similar example showing that even when the closed loop pole positions differ greatly, the overall response of the system does not necessarily change much.

2.9. The S Domain.

While the CACSD package has been designed for z domain transfer functions, the modifications necessary for it to operate in the s domain should not be difficult to effect. Frequency domain responses will be evaluated at points on the imaginary axis instead of the unit circle; time domain responses will require integration of differential equations instead of summation of difference equations. Possibly the most important change is that the elements of the **Q** matrices will require a different parameterization; in general they will still take the form of a transfer function where the denominator coefficients are fixed, and the numerator coefficients are computed by the quadratic programming algorithm. It may also be feasible to transfer the FIR filter structure of these elements in the z domain to the s domain using a sum of e^{-skT} terms with variable coefficients. However the z domain transfer functions remain better suited for digital computer implementation, and the algorithms for that domain are usually more efficient than their corresponding s domain counterparts.

2.10. Summary.

To improve the efficiency of the design method of Boyd *et al.* [30], a diagonal factorization technique has been developed. This allows the multivariable design problem to be reduced to a number of smaller sub-problems, which may then be solved independently. Although the diagonal factorization is not always coprime, it is suitable for a wide range of plant transfer function matrices. A theorem to check that the factorization is coprime has been developed, and is easy to apply. Formulae for (non-diagonal) coprime factorizations, where the nominal stabilizing controller is stable, have also been presented. A novel parameterization for the design transfer function **Q** has been introduced; by appropriate choice of the QSTEP parameter the designer benefits from the efficiency of a low order approximation and while in many cases enjoying almost the same precision as for a high order approximation.

CHAPTER 3
IMPLEMENTATION OF THE CACSD PACKAGE

3.1. Introduction.

This chapter describes the implementation of a CACSD package based on the design method outlined in chapter 2. In addition to the design algorithm, the package includes a number of ancillary components to provide a collection of resources upon which the supervising expert system can call. The main functions of this package are shown in figure 3.1, and the interface between it and the expert system is given in appendix B.

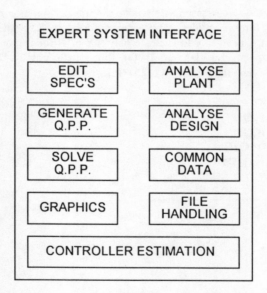

Figure 3.1. Overall structure of the CACSD package.

The bulk of this chapter deals with the generation and solution of the quadratic programming problem. First the translation of the system performance constraints into linear constraints on the search vector is discussed; the translation of the quadratic cost function is treated here also. Next a representation for the efficient storage and manipulation of these constraints is described. Finally the algorithm used to solve the quadratic programming problem is examined.

The supporting functions provided by the package are also briefly described at the end of this chapter.

3.2. Generating the Quadratic Programming Problem.

In order to translate the control system design problem into a quadratic programming problem, it is necessary to compute the matrix fractions of the plant and initial stabilizing controller. Following that, the performance constraints (in both the time and frequency domains) can be translated into linear constraints on the search (parameter) vector; the cost function can similarly be translated into a function of the search vector. These steps are examined below.

3.2.1. Computing the Matrix Fractions.

The closed loop transfer functions $H_c(z)$ are evaluated using the equation

$$H_c = H0_c + H1_c Q H2_c \tag{3.1}$$

However it is not always necessary to determine the transfer function matrices $H0_c$, $H1_c$, and $H2_c$ in algebraic form. Considering responses from inputs R, N and D, it can be seen from table 3.1 that it is necessary to evaluate only **NX** or **N** for responses with output Y ($H0_Y$ and $H1_Y$ respectively), **DX** or **D** for those with output U ($H0_U$ and $H1_Y$ respectively), and $\tilde{D}$ when the input is N or D (**H2**).

Table 3.1. Closed loop transfer functions.

H_{RY} =	0	+ $H1_Y Q1$
H_{RU} =	0	+ $H1_U Q1$
H_{NY} =	$H0_Y$	+ $H1_Y Q2 H2$
H_{NU} =	$H0_U$	+ $H1_U Q2 H2$
H_{DY} =	I - H_{NY}	
H_{DU} =	-H_{NU}	

In table 3.1 above,

$H0_Y = NX$,
$H1_Y = N$,
$H0_U = DX$,
$H1_U = D$,
and $H2 = \tilde{D}$.

Table 3.2 below gives the forms used for **H0** and **H1** depending on the stability of the plant and nominal controller, and whether or not a right-coprime factorization of the plant is available. Note that in the case where right factorization of the plant is not coprime, and the controller is not stable, the design will be sub-optimal. This is also the case when the left factorization of the plant is not coprime. In this table,

Chapter 3. Implementation of the CACSD Package. 27

	G_{stab}	indicates whether the plant is stable,
	K	indicates whether a nominal controller **K0** has been specified,
	G_{co}	indicates whether the diagonal right factorization $G = ND^{-1}$ is coprime,
	K_{stab}	indicates whether the nominal controller is stable,
	F	is the common factor $(I + K_0G)^{-1}$
and	x	means yes or no.

Table 3.2. Formulae for computing H0 and H1.

G_{stab}	K	G_{co}	K_{stab}	$H0_U$	$H0_Y$	$H1_U$	$H1_Y$
yes	no	x	x	0	0	I	G
yes	yes	x	x	FK0	GFK0	I	G
no	yes	no	yes	FK0	GFK0	F	GF
no	yes	yes	x	FK0	GFK0	D	N
no	yes	no	no	FK0	GFK0	D	N
no	no	x	x		NOT ALLOWED		

Frequency responses are evaluated at a set of discrete points, logarithmically spaced on the unit circle in the complex plane. The matrices **H0**, **H1** and **H2** are evaluated at these frequencies, and stored for future use. For step responses, only the impulse responses of **H0**, **H1** and **H2** must be stored. Impulse responses for the expressions in table 3.2 involving the complicated common factor **F** are evaluated through a simulation of the corresponding closed loop system made up of **G** and **K0**. Using the notation of figure 3.2, **FK0** represents the closed loop response from input N to output U, **GFK0** that from N to Y, **F** that from V to U, and **GF** that from V to Y. Thus the algebraic forms of these matrices need not be determined explicitly. The package computes and stores these impulse and frequency responses; thus they need not be recomputed each time a response is evaluated, which saves much time.

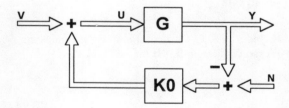

Figure 3.2. Closed loop system used to evaluate impulse responses.

3.2.2. Translating the Performance Constraints.

Performance constraints on the closed loop responses are translated into linear constraints on the search vector, as shown by Boyd [30]. Given the diagonal factorization of chapter 2, the equation for the closed loop response of element [i,j] can be written as

$$\mathbf{H}_c[i,j] = \mathbf{H0}_c[i,j] + \sum_{k=1}^{n} \mathbf{H3}_c[i,k]_j \mathbf{Q}[k,j] \tag{3.2}$$

where

$$\mathbf{H3}_c[i,k]_j = \mathbf{H1}_c[i,k]\mathbf{H2}_c[j,j]. \tag{3.3}$$

For clarity, the analysis below assumes a value of one for the QSTEP parameter used to model the elements of $\mathbf{Q}$. This makes $\mathbf{Q}[i,j]$ a standard p-tap FIR filter, which is parameterized as

$$\mathbf{Q}[i,j] = \sum_{r=0}^{p-1} q_{i,j,r} z^{-r} \tag{3.4}$$

where $q_{i,j,r} \in \mathbf{R}$. Now define a parameter vector $\mathbf{x}_j \in \mathbf{R}^{np}$ made up of the coefficients of the FIR filters in column j of $\mathbf{Q}$, according to the formula

$$q_{i,j,r} = \mathbf{x}_j[(i-1)n + r]. \tag{3.5}$$

Then the closed loop step response of a particular element at time uT may be written as

$$\mathbf{h}_c[i,j](uT) = \mathbf{h0}_c[i,j](uT) + \sum_{k=1}^{n}\sum_{r=0}^{p-1} \mathbf{h3}_c[i,k]_j((u-r)T)q_{k,j,r}$$

$$= d0 + \sum_{k=1}^{n}\sum_{r=0}^{p-1} d_{i,j,k,r} q_{k,j,r}$$

$$= d0 + \mathbf{d}^T \mathbf{x}_j \tag{3.6}$$

where $d0, d_{i,j,k,r} \in \mathbf{R}$, and $\mathbf{d} \in \mathbf{R}^{np}$. Thus constraints on the minimum or maximum value of a closed loop step response element [i,j] at time uT

$$v1 \leq \mathbf{h}_c[i,j](uT) \leq v2 \tag{3.7}$$

can be transformed into linear constraints on the parameter vector $\mathbf{x}$

$$v1 - d0 \leq \mathbf{d}^T \mathbf{x}_j \leq v2 - d0. \tag{3.8}$$

The closed loop frequency response of a particular element, evaluated at a specific point ω on the unit circle, may be written as

Chapter 3. Implementation of the CACSD Package.

$$\mathbf{H}_c[i,j](\omega) = \mathbf{H0}_c[i,j](\omega) + \sum_{k=1}^{n}\sum_{r=0}^{p-1}\mathbf{H3}_c[i,k]_j(\omega)\omega^{-r}q_{k,j,r}$$

$$= c0 + \sum_{k=1}^{n}\sum_{r=0}^{p-1}c_{i,j,k,r}q_{k,j,r}$$

$$= c0 + \mathbf{c}^T\mathbf{x}_j \tag{3.9}$$

where $c0$, $c_{i,j,k,r} \in \mathbf{C}$, and $\mathbf{c} \in \mathbf{C}^{np}$.

A constraint on the maximum modulus of this closed loop frequency response

$$|\mathbf{H}_c[i,j](\omega)| \leq m, \tag{3.10}$$

$m \in \mathbf{R}$, $m \geq 0$, then becomes

$$|c0 + \mathbf{c}^T\mathbf{x}_j| \leq m. \tag{3.11}$$

Splitting these into real and imaginary components $c0_r$, $c0_i \in \mathbf{R}$, and $\mathbf{c}_r$, $\mathbf{c}_i \in \mathbf{R}^{np}$, gives

$$(c0_r + \mathbf{c}_r^T\mathbf{x}_j)^2 + (c0_i + \mathbf{c}_i^T\mathbf{x}_j)^2 \leq m^2. \tag{3.12}$$

Boyd *et al.* [30] have shown how this complex modulus constraint may be approximated by linear constraints; an efficient representation [48] developed for these is discussed later. Using these techniques, the constraint on the modulus of the frequency response element [i,j] may also be translated into linear constraints on the parameter vector **x**.

The linear constraints generated are grouped according to the performance constraints they represent; these groups are presented to the quadratic programming algorithm until all have been satisfied, or a conflict is found. When there is no feasible solution, corresponding to a conflict in the design specifications, the expert system uses the status of each group (satisfied, active, not satisfied, or not yet attempted) to explain the cause of the conflict (see section 5.5.6).

3.2.3. Translating the Cost Function.

The design method also allows for optimization of the closed loop responses. The quadratic cost function to be minimized is again based on these responses; the cost is defined as the square of the difference between the response and some constant offset. In the time domain, the response of are particular element, at time uT, has a cost function

$$J_{i,j} = (\mathbf{h}_c[i,j](uT) - e)^2$$

$$= (\mathbf{d}^T\mathbf{x}_j + d0 - e)^2$$

$$= \mathbf{x}_j^T(\mathbf{dd}^T)\mathbf{x}_j + 2(d0 - e)\mathbf{d}^T\mathbf{x}_j + (d0 - e)^2 \tag{3.13}$$

associated with it, where $e \in \mathbf{R}$ is an constant offset (usually 0 or 1). In the frequency domain, at frequency ω, the corresponding cost function is

$$\begin{aligned}
J_{i,j} &= |\mathbf{H}_c[i,j](\omega) - f|^2 \\
&= (c0_r + \mathbf{c}_r^T \mathbf{x}_j - f_r)^2 + (c0_i + \mathbf{c}_i^T \mathbf{x}_j - f_i)^2 \\
&= \mathbf{x}_j^T(\mathbf{c}_r \mathbf{c}_r^T + \mathbf{c}_i \mathbf{c}_i^T)\mathbf{x}_j + 2(c0_r - f_r)\mathbf{c}_r^T \mathbf{x}_j + 2(c0_i - f_i)\mathbf{c}_i^T \mathbf{x}_j \\
&\quad + (c0_r - f_r)^2 + (c0_i - f_i)^2
\end{aligned} \quad (3.14)$$

where $f_r, f_i \in \mathbf{R}$ are the real and imaginary components of the offset constant $f \in \mathbf{C}$ (also usually 0 or 1). Note that in both cases the cost function is usually scaled by a real positive constant, referred to as the optimization weight. Furthermore, the symmetric matrices $\mathbf{dd}^T$, $\mathbf{c}_r\mathbf{c}_r^T$, and $\mathbf{c}_i\mathbf{c}_i^T$ are positive semi-definite; in practice the sum of many of these terms is generally positive definite. Thus there is always some finite value of $\mathbf{x}_j$ giving the minimum cost, and this value is almost always unique.

3.3. Representing the Linear Constraints[2].

An efficient representation for the linear constraints generated by the design method has been developed by the author [48]. This representation exploits the inherent structure of these constraints to reduce both the storage requirements and execution time of the quadratic programming algorithm substantially, without changing the characteristics of this algorithm in any way. The gains in efficiency are particularly significant for the frequency domain constraints. In the examples given, the constraints are represented using only about 15% of storage space for the equivalent simple linear constraints; execution speed was increased by a factor between 1.7 and 5.1 at the same time.

Formulae developed for the compound representation, which treats groups of linear constraints simultaneously, turn out to be much more efficient than the corresponding formulae for the equivalent group of linear constraints in standard form. Most of the speed advantage is achieved at lines 45-56 in the QPSOL algorithm listed in section 3.5; during phase I some advantage is also obtained at lines 9-14.

The Gill-Murray active set algorithm used to solve the quadratic programming problem performs two distinct operations on the individual constraints. The first operation determines if a constraint has been satisfied; this is computed for each previously unsatisfied constraint during every iteration of phase I. The second operation computes the distance, in a specified search direction, to the constraint boundary; this must be computed for each non-active constraint during at least half of the total number of iterations. For problems with a large number of constraints and a relatively small number of decision variables, which is typical for this application, these two activities are the most time-consuming sections of the algorithm. The representation discussed below relates to these two activities alone.

---

[2]This section (c) IEEE, and is reprinted, with permission, from IEEE Transactions on Automatic Control, vol. 35, no. 8, pp. 949-951, 1990.

In the present implementation, the constraints are stored as a list of records. Each record is denoted by [a1,a2,b,T], where $a1 \in \mathbf{R}$, $a2 \in \mathbf{R}$, and $\mathbf{b} \in \mathbf{R}^n$. T defines the constraint type.

3.3.1. Time Domain Constraints.

Time domain constraints on a closed loop step response h(t) are evaluated at discrete sampling intervals. At a specific time t_1, the constraints may be posed in the form

$$k1 \leq h(t_1) \leq k2 \tag{3.15}$$

where $k1 \in \mathbf{R}$ and $k2 \in \mathbf{R}$.

In some cases only one bound is present, or the bounds are equal (an equality constraint). Here the resulting linear constraints are in standard form, and neither require nor benefit from any special representation. The remainder of the section will deal with the case where both bounds are present and distinct.

The step response, evaluated at time t_1, can be written in terms of the parameter vector $\mathbf{x} \in \mathbf{R}^n$, as

$$h(t_1) = a + \mathbf{b}^T\mathbf{x} \tag{3.16}$$

where $a \in \mathbf{R}$ and $\mathbf{b} \in \mathbf{R}^n$. This results in the linear constraints

$$k1 - a \leq \mathbf{b}^T\mathbf{x} \leq k2 - a \tag{3.17}$$

which can be stored as the record

[k1-a, k2-a, **b**, between]

Combining both constraints into one record also gives a computational advantage since only one instead of two dot products need be computed to evaluate the pair. When computing the distance to the constraint boundary, there are additional advantages :

- If the lower (respectively upper) bound of the constraint is active, then the upper (respectively lower) bound need not be checked as it cannot be violated.

- The sign of the scalar product of the search direction with the constraint vector constant indicates directly which bound (upper or lower) needs to be checked; if the scalar product is zero then neither bound is significant during that particular iteration.

3.3.2. Frequency Domain Constraints.

Frequency domain constraints on a closed loop transfer function $H(\omega)$ are approximated by evaluating H at discrete frequencies. At a specific frequency ω_1 the constraint is in the form

$$|H(\omega_1)| \leq k \tag{3.18}$$

where $k \in \mathbf{R}$ and $k > 0$.

This complex modulus constraint is not a linear constraint; in the complex plane the constraint boundary is a circle of radius k. However the circle may be approximated by linear constraints as shown in [30]. The present implementation approximates the circle using 16 linear constraints, resulting in a maximum approximation error of less than one percent. For this choice the boundary of the feasible region falls between an inner circle of radius $k\sqrt{\alpha}$ and an outer of radius $k/\sqrt{\alpha}$, where

$$\alpha = \cos(\pi/16) \approx 0.98 \tag{3.19}$$

Figure 3.3 illustrates the geometry of the approximation; for the sake of clarity, the complex modulus constraint is approximated by eight linear constraints, and only one quadrant is shown.

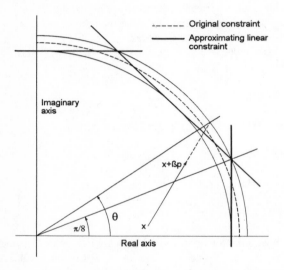

Figure 3.3. Approximation of a complex modulus constraint using linear constraints.

The transfer function evaluated at a frequency ω_1 can be written in terms of the parameter vector $\mathbf{x}$ as

$$H(\omega_1) = a + \mathbf{b}^T\mathbf{x} \tag{3.20}$$

where $a \in \mathbf{C}$ and $\mathbf{b} \in \mathbf{C}^n$.

When split into separate real and imaginary components as

$$\mathrm{Re}(H(\omega)) = c + \mathbf{d}^T\mathbf{x}$$

and

$$\mathrm{Im}(H(\omega)) = e + \mathbf{f}^T\mathbf{x} \tag{3.21}$$

with $c \in \mathbf{R}$, $\mathbf{d} \in \mathbf{R}^n$, $e \in \mathbf{R}$, and $\mathbf{f} \in \mathbf{R}^n$, the complex modulus constraint may be written as

Chapter 3. Implementation of the CACSD Package.

$$(c + \mathbf{d}^T\mathbf{x})^2 + (e + \mathbf{f}^T\mathbf{x})^2 \leq k^2 \tag{3.22}$$

and stored as the pair of records

$$[c, k, \mathbf{d}, \text{complex\_1}], \quad [e, k, \mathbf{f}, \text{complex\_2}]$$

This gives a great storage advantage over storing 16 such elements; if required any of the original 16 may be easily recovered as a linear combination of these two. As for the time domain case there are computational advantages too; for evaluation purposes only two dot products need to be computed, instead of 16. This representation does require a small overhead in tracking which of the 16 combinations are active.

The discussion below relates to a single iteration of the QPSOL algorithm. The current point denotes the projection of the current parameter vector $\mathbf{x}$ onto the complex plane, and the search direction the projection of the current search vector $\mathbf{p}$. A distance β in this direction is computed in terms of the vector equation

$$\mathbf{x}' = \mathbf{x} + \beta\mathbf{p} \tag{3.23}$$

Using the geometry of the line in the search direction through the current point, the distance from the current point to the inner circle (β_i), and to the outer circle (β_o), are easily computed as

$$\beta_i = -g + \sqrt{g^2 + h_i} \tag{3.24}$$

$$\beta_o = -g + \sqrt{g^2 + h_o} \tag{3.25}$$

where

$$g = \frac{(c + \mathbf{d}^T\mathbf{x})(\mathbf{d}^T\mathbf{p}) + (e + \mathbf{f}^T\mathbf{x})(\mathbf{f}^T\mathbf{p})}{(\mathbf{d}^T\mathbf{p})^2 + (\mathbf{f}^T\mathbf{p})^2} \tag{3.26}$$

$$h_i = \frac{k^2\alpha - (c + \mathbf{d}^T\mathbf{x})^2 - (e + \mathbf{f}^T\mathbf{x})^2}{(\mathbf{d}^T\mathbf{p})^2 + (\mathbf{f}^T\mathbf{p})^2} \tag{3.27}$$

$$h_o = \frac{k^2/\alpha - (c + \mathbf{d}^T\mathbf{x})^2 - (e + \mathbf{f}^T\mathbf{x})^2}{(\mathbf{d}^T\mathbf{p})^2 + (\mathbf{f}^T\mathbf{p})^2} \tag{3.28}$$

Further advantages of this representation are derived as follows

- If the condition

$$(c + \mathbf{d}^T\mathbf{x})^2 + (e + \mathbf{f}^T\mathbf{x})^2 < k^2\alpha \tag{3.29}$$

is satisfied (the current point is inside the inner circle), then all 16 approximating linear constraints are satisfied and inactive.

- At most 2 of the 16 approximating linear constraints may be active at any time. In addition, if one of these constraints is active, then only the adjacent (in a circular sense) linear constraints need to be checked; the distance to the relevant constraint is β_o. The remaining 14 constraints are also known to be satisfied if two adjacent constraints are active.

- When β_i is real, the distance in the search direction to the nearest of the 16 approximating linear constraints is at least β_i. β_i is real if the line in the search direction through the current point intersects the inner circle. If β_i is larger than the current minimum distance to a constraint boundary, then no further checks are required on this complex constraint during the present iteration. Failing this, the particular linear constraint (1 of the 16) which needs to be checked explicitly can be determined from the point of interception of the line in the search direction through the current point, and the outer circle. The formula for the angle of this point in the complex plane, which determines the relevant constraint, is

$$\theta = \tan^{-1}\left[\frac{(e+\mathbf{f}^T\mathbf{x})+\beta_o(\mathbf{f}^T\mathbf{p})}{(c+\mathbf{d}^T\mathbf{x})+\beta_o(\mathbf{d}^T\mathbf{p})}\right] \tag{3.30}$$

- When β_i is complex, it is possible to determine which of the 16 constraints is the nearest to the current point, in terms of the search direction. This involves computing the tangents to the outer circle which pass through the current point. An alternative is to use these tangents as additional constraints; since they are tangent to the outer circle, they do not change the feasible set. Clearly the distance to either tangent is zero. Neither of these two features have been implemented in the modified quadratic programming algorithm, as the computations involved are sufficiently complex to outweigh their advantages. However should the number of lines approximating the circle be substantially larger than 16, these computations will be beneficial.

3.3.3. Examples.
Two control system design problems are used to illustrate the relative efficiency of the representation described above.

Example 1.
The first is the design of a two parameter controller for a SISO plant, using a parameter vector of order 5. The plant transfer function, sampled at a rate of 1 Hz, is

$$G(z) = \frac{0.15z^{-4}}{z-1}$$

The precompensator (**K1**) was designed using time domain constraints on the response of

Chapter 3. Implementation of the CACSD Package.

the plant output y(kT) and the control signal u(kT) to a unit step at the reference input; the constraints set were

$$0 \leq y(kT) \leq 1.05, \quad 0 \leq kT < 20s,$$
$$0.9 \leq y(kT) \leq 1.05, \quad 20 \leq kT < 30s,$$
$$0.95 \leq y(kT) \leq 1.05, \quad 30 \leq kT \leq 50s,$$
and $\quad -2.0 \leq u(kT) \leq 2.0, \quad 0 \leq kT \leq 50s.$

A quadratic cost function based on the plant output error was specified, and the constrained optimal solution found after 9 iterations. The standard quadratic programming algorithm required 194 linear constraints, and was solved in 26.9 seconds on an 8088 based personal computer. The modified program required 97 constraint records, and the solution was found in 15.8 seconds.

The feedback element (**K2**) was designed using frequency domain constraints on the closed loop gain $H_{NY}(\omega)$, and the output disturbance response $H_{DY}(\omega)$, as shown below.

$$|H_{NY}(e^{j2\pi f})| \leq 1.2, \quad 0 \leq f \leq 0.5 \text{ Hz}$$
and
$$|H_{DY}(e^{j2\pi f})| \leq 0.1, \quad 0 \leq f \leq 0.004 \text{ Hz}$$

These constraints were evaluated at 25 discrete frequencies. The optimal solution was found after 17 iterations, given a quadratic cost function based on $|H_{NY}(\omega)|^2$. The standard algorithm found the solution in 124.5 seconds, using 548 linear constraints. The modified program used 72 constraint records, and took 35.2 seconds.

Example 2.

The second example is the design of a feedback controller (**K2** only) for a MIMO plant with the transfer function

$$G(z) = \begin{bmatrix} \dfrac{0.1}{z-0.9} & \dfrac{0.3}{z-0.9} \\ \dfrac{0.2}{z-0.9} & \dfrac{0.8}{z-0.8} \end{bmatrix}$$

Given a sampling rate of 1 Hz, the closed loop system is required to meet the constraints

$$H_{DY}[i,j](1) = 0$$
$$|H_{DY}[i,j](e^{j2\pi f})| \leq 0.25 \quad 0 \leq f \leq 0.5 \text{ Hz}, i \neq j$$
$$|H_{DY}[i,j](e^{j2\pi f})| \leq 1.5 \quad 0 \leq f \leq 0.5 \text{ Hz}$$
$$|h_{DU}[i,j](kT)| \leq 5.0 \quad 0 \leq kT \leq 100 \text{ seconds}$$

for $i,j \in \{1,2\}$, and minimize the cost function

$$J = \sum_{k=0}^{100} \left[0.1 h_{DY}^2[1,1](kT) + h_{DY}^2[1,2](kT) + h_{DY}^2[2,1](kT) + 0.1 h_{DY}^2[2,2](kT) \right]$$

Using the diagonal factorization technique the design was split into two sub-problems which were then solved independently. For each of these a parameter vector of order 10 was used, and the frequency domain constraints were evaluated at 100 discrete points on the unit circle. The sub-problems were solved in 38 and 55 iterations respectively.

For the standard algorithm each sub-problem translated into 3406 linear constraints; the optimum solutions were found in 20.1 and 32.2 seconds respectively using an 80386/387 based computer. The modified algorithm required only 504 constraint records for each sub-problem, and produced the solutions in 5.3 and 6.3 seconds respectively.

3.4. Solving the Quadratic Programming Problem.

The function of the quadratic programming algorithm QPSOL is to find the vector **x** which minimizes the quadratic cost function

$$\tfrac{1}{2} x^T E x + e^T x, \tag{3.31}$$

and satisfies the set of linear constraints

$$C(x) \geq 0 \tag{3.32}$$

Each constraint C_i is represented by the linear equation

$$C_i(x) = a_i^T x + b_i. \tag{3.33}$$

A two-phase algorithm is used to solve the quadratic programming problem; phase I locates a feasible point and, if one exists, phase II then finds the constrained optimum solution. This algorithm is based on two of the Gill-Murray active set methods given in Scales [49].

For clarity of the algorithm shown below, it is assumed that there is at least one constraint. If not, phase I of the algorithm is omitted, and the problem reduces to solving

$$Ex + e = 0 \tag{3.34}$$

for **x**.

The columns of the active set **A** are made up of the vectors a_i of those constraints C_i which are active. An orthogonal QR factorization of **A**,

$$A = QR, \tag{3.35}$$

with **Q** column orthonormal ($Q^T Q = I$), and **R** upper triangular, is used extensively in the algorithm. This **Q** matrix should not be confused with the matrices **Q1** and **Q2** used to

Chapter 3. Implementation of the CACSD Package.

parameterize the closed loop transfer functions. Scales [49] gives efficient techniques for updating these QR factors when a column is added to or deleted from **A**, based upon the Householder transformation [50]. The computational efficiency of the factorization has been significantly improved by taking advantage of the structure of matrix **R** and the Householder transformation matrix **P** during matrix multiplications.

When q constraints are active, the QR factors can be partitioned as

$$\mathbf{A} = \begin{bmatrix} \mathbf{Qa} & \mathbf{Qb} \end{bmatrix} \begin{bmatrix} \mathbf{R} \\ \mathbf{0} \end{bmatrix} \qquad (3.36)$$

where **Qa** has q columns and **R** has q rows. The matrix **Qb**, with np-q columns, is used frequently in the QPSOL algorithm.

3.5. The QPSOL Algorithm.

```
1    Set phase2 = feasible = terminate = try_del = false
2    Set α = 1, q = 0
3    Set x = 0, A = 0, Q = I
4    REPEAT
5        IF (α > 0)
6            IF (phase2)
7                Set g = Ex + e
8            ELSE
9                IF (C_i(x) ≥ 0 for all i)
10                   Set feasible = terminate = true
11               ELSE
12                   Set g = - Σa_i for all i
13                   such that C_i(x) < 0
14               END
15           END
16       END
17       Set deletion = false
18       IF (NOT terminate)
19           IF (NOT try_del)
20               Set v = Qb^T g
21               IF (phase2)
22                   Solve Qb^T E.Qb.y = v for y
23                   Set v = y
24               END
25               Set p = -Qb.v
26           END
27           IF (try_del OR (|Qb^T g| < eps))
28               Solve Rμ = Qa^T g for μ,
29               the Lagrange multipliers.
30               IF (no μ_i < 0)
31                   Set terminate = true
32               ELSE
33                   Find smallest μ_i, and delete the
34                   corresponding constraint from A
35                   Set q = q - 1
36                   Update the factorization A = Q.R
37                   Set deletion = true
```

```
38                Set α = 0
39            END
40        END
41    END
42    Set try_del = false
43    IF (NOT(terminate OR deletion))
44        Set addition = true
45        Find the smallest distance β in direction p
46        to any constraint (C_j) which is currently
47        satisfied but not active.
48        IF (phase2)
49            β = 1.001
50        ELSE
51            IF (none found in this direction)
52                Find the distance β in direction p to
53                the furthest constraint (C_j) which is
54                currently violated.
55                Set addition = false
56            END
57        END
58        IF (phase2 AND (β > 0.999))
59            Compute the distance δ to the minimum of the
60            cost function in the direction p, using the
61            formula
62                    δ = -(p^T E p)/(g^T p)
63            IF (δ < β)
64                Set β = δ
65                Set addition = false
66                Set try_del = true
67            END
68        END
69        Set α = β
70        IF (α > 0)
71            Set x = x + αp
72        END
73        IF (addition)
74            Add constraint C_j to active set A
75            Update factorization A = Q.R
76            Set q = q + 1
77        ELSE
78            IF ((NOT phase2) OR (α < eps))
79                Set terminate = feasible = true
80            END
81        END
82    END
83    IF (feasible AND (NOT phase2))
84        Set terminate = false, phase2 = true
85        Set α = 1
86    END
87 UNTIL (terminate)
88 END
```

During phase II, the section of the algorithm for computing the smallest distance to a constraint boundary (line 45), takes advantage of the implicit scaling of the algorithm. This scaling is such that the distance β to the minimum of the cost function, subject to the active constraints, is unity. Rounding errors (even when using double precision arithmetic)

Chapter 3. Implementation of the CACSD Package. 39

make this distance only approximately one, and the formula at line 62 is found to give a more accurate assessment of the distance.

Having solved the quadratic programming problem for **Q1** and **Q2**, it is possible to compute the controller transfer functions **K1** and **K2** from the formulae (2.4). This calculation is not performed explicitly (in transfer function form); rather **Q1** and **Q2** are used directly to implement the controller, or to estimate a low order approximation, as discussed in chapter 8.

3.6. Ancillary Functions of the CACSD Package.

3.6.1. Entry of Transfer Function Matrices.

The plant transfer function matrix $G(z)$ is read from a file, or entered into a spreadsheet style form. Left and right stable diagonal matrix fractions of the plant are then computed; the package also checks that they are coprime, and returns this information to the expert system. Under certain circumstances, for example when the plant is unstable, a nominal stabilizing controller transfer function matrix $K(z)$ must also be specified.

3.6.2. Entry of Performance Constraints.

Performance constraints on individual closed loop step and frequency responses, as well as those on the singular values, are entered using the form shown in figure 3.4. The expert system discussed in chapters 4 and 5 also allows constraints to be entered on multiple elements of a response simultaneously, by copying the constraints entered on one element of a response to other elements as required.

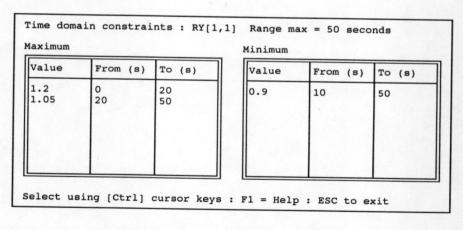

Figure 3.4. Entry of performance constraints.

3.6.3. Graphics Facilities.

The package includes graphics facilities for plotting the system step and frequency responses, as well as the minimum and maximum singular values [51] of the various frequency responses; performance constraints on the particular response are also shown on the plot. The direct and inverse Nyquist arrays, with Gershgorin circles, may also be plotted.

3.6.4. Analysis Facilities.
Some of the decisions taken by the expert system to guide the user are based on analyses of the design performed by the CACSD package. The CACSD package can, for example, compute the actual closed loop response over a time or frequency interval, or determine whether or not a specification on the singular values of a frequency response has been satisfied.

3.6.5. Reduced Order Controller Estimation.
The CACSD package also includes a simple controller estimation facility which allows a reduced order controller to be estimated and then evaluated. Details of this procedure are given in chapter 8.

3.7. Summary.

This chapter has dealt with issues relating to the implementation of the CACSD package. Efficient forms for evaluating the coprime matrix fractions have been given, and the translation of the control system design problem into a quadratic programming problem has been described. A very efficient representation for the linear constraints generated by this method, both in terms of memory requirements and execution speed, has been developed. Finally an algorithm for solving the resulting quadratic programming problem, derived from two active-set methods, is given.

CHAPTER 4
THE EXPERT SYSTEM

4.1. Introduction.

Although the numerical design method discussed in chapters two and three is extremely powerful, it forms only one step in the overall design process. There are many issues which need to be addressed in order to apply it successfully, and it is here where the expert system proves invaluable. Broadly speaking, the expert system coordinates the functions of the CACSD package, and provides an intelligent user interface. It assists the designer in formulating a comprehensive and achievable design specification, and in dealing with conflicting design constraints. The expert system also effectively extends the scope of the underlying CACSD package.

The CACSD package requires that the design be specified in terms of performance constraints on the closed loop response of the system. The design cycle comprises setting performance constraints and a cost function for the control system, and then using the CACSD package to find a controller which satisfies these. In general this is an iterative process whereby the specification is engineered in stages, with specifications being added, tightened, or relaxed at each step, until satisfactory performance is achieved. The main function of the expert system is to assist the user in formulating and refining this specification; the assistance takes place at many levels, ranging from simply explaining the commands available and how they should be used, to providing advice on how the specification could be improved.

The designer interacts with the design system using commands; these commands may be divided into two main groups. The first group interacts directly with the CACSD package, and comprises those commands used for editing the specification, solving for a controller, and examining the performance of the controller. For example, the EDIT, SOLVE and PLOT commands fall into this category. The second comprises those commands used to assist the designer, and includes the HELP, SUGGEST and NEXT STEP commands. This second group, and the most important tasks performed by the expert system, are examined in detail below[3].

---

[3]Parts of this chapter and the next are reprinted from Automatica, vol. 28, no. 3, C.D. Tebbutt, An Expert System for Multivariable Controller Design, pp. 463-471, Copyright 1992, with kind permission from Elsevier Science Ltd, The Boulevard, Langford Lane, Kidlington OX5 1GB, UK.

4.2. The User Interface and Philosophy.

The expert system attends to almost all of the interaction with designer. Unfortunately novice and experienced designers have different requirements of a user interface. Experienced designers need ready access to comprehensive and detailed information on the state of the design, and complete freedom to choose from all possible design alternatives; the expert system should also contribute towards improving the productivity of the expert. Novice designers are often overwhelmed by too much information, or too many degrees of freedom in the design method; they need assistance in interpreting the information, and guidance in choosing amongst the design options. Infrequent users of a design system, novice or otherwise, may require help with the design language. Taylor and Frederick [12] give further examples of areas where "less-than-expert" users experience difficulty with traditional CACSD software. The expert system user interface aims to deal with these issues.

The interface is largely command-driven for ease of use by proficient users; the expert system gives the experienced designer almost direct access to the underlying CACSD package, with only brief comments on the design progress. To assist novice and infrequent users there are also commands such as HELP, SUGGEST, and NEXT STEP. The concept of an expert system intercepting these types of commands has been termed "command spy" by Larsson and Persson [4]. This structure allows users to request advice and assistance when desired. As the user's knowledge and experience with the system grows, they will tend to use the advice facilities less frequently.

Pang and MacFarlane [10] propose that matters of engineering judgment be left to the designer. This application follows the same philosophy, with the responsibility for all the final design decisions resting on the designer. The expert system has little knowledge of the physical system to be controlled; for example it does not know whether a high frequency gain of +10 dB on the NU response is excessive or not. While it can and does generate warnings and reminders, this type of decision is ultimately left to the designer.

On the whole, the expert system does not question the specifications placed by the designer; it is assumed that the designer has some constructive purpose for each specification. However there are certain situations where the designer is warned of suspicious specifications. Similarly, the expert system does little in the way of checking that the advice given to the user is applied; it is assumed that the designer may have a valid reason for ignoring it.

The expert system has no automatic learning capabilities. Nevertheless, after many design sessions and through the skill of the knowledge engineer, the expert system has undergone a substantial indirect "learning" experience, resulting in the current version of the design tool being significantly more capable than earlier versions [52,53]. One of the strengths of the expert systems approach is the relative ease with which this learning, or extension of the knowledge base, can be effected [12].

4.3. Explaining the Design Language and Methodology.

The design language, terminology and the overall design philosophy for the CACSD design method may be unfamiliar to novice or infrequent users; MacFarlane *et al.* [29] point out that expert systems may contribute towards making design packages more easily

Chapter 4. The Expert System.

accessible in these situations. The expert system contains a large amount of information on the use of the design tool, covering the design language and notation, the syntax and use of the commands, and the general philosophy for using this design method. While much of this is a simple menu-driven help system as shown below, or is programmed into the expert system explanation facility, a small measure of reasoning is used in generating the context-sensitive help. One of the objectives of the interface is to tutor the user in applying the design tool effectively.

```
DY[1,1](frequency) >HELP
Help is available in the following areas :

1.      The design language and philosophy.
2.      A particular command.
3.      The step-by-step design guide.
4.      Refining your design.
5.      Dealing with conflicting constraints.
6.      Checking that the design is complete.

Which of these do you need help with ? 1

Choose one of the following :

1.      The design method overview.
2.      Terminology used in this program.
3.      How to plan your design.
4.      Selecting a response to work on.
5.      Plotting a step or frequency response.
6.      Setting constraints on a response.
7.      Setting up an optimization function.
8.      Finding a suitable controller.
9.      Selecting the design parameters.

....
```

Genesereth [54] shows how an intelligent system may reconstruct the user's plan of action in order to provide advice on the correct usage of Macsyma, a symbolic mathematics program. Jackson and Lefrere [55] develop this concept further. Larsson and Persson [4] store typical command sequences in scripts, and match these against the history of commands issued by the user in an attempt to understand the user's intentions. Similar types of analyses could also be used to assist the control system designer. For example, when an invalid command is issued, the expert system could attempt to identify the designer's intentions from the history of design steps, and then explain the correct usage of the commands. At present it is felt that these approaches are not justified, in light of the commands seldom being chained into lengthy sequences but rather being used somewhat independently. A simple notification of the error, combined with suggestions on how to obtain further help, suffices.

Nevertheless the expert system does monitor how it is being used by the designer. This information is useful, for example, when deciding whether or not certain features of the design system deserve explanation.

4.4. Presentation of the Design Status.

One of the most important functions of any design system is that of providing the designer with information on the status of the design; this information should be both comprehensive and easily comprehended.

The CACSD package on its own does not provide comprehensive information on the design in the sense that only information relating to the specifications is made available, and the range of specifications permitted is limited. The expert system, often using supporting numerical software in the CACSD package, attempts to effectively increase the range of specifications available, and also performs further analyses on the design, supplementing the information available. This aspect of the expert system is dealt with in detail later.

The expert system also attempts to provide the design information in an easily comprehended form. An example of this is the analysis of the Lagrange multipliers from the quadratic programming algorithm. From these it is possible to determine which of the high level performance constraints were satisfied, active or could not be satisfied; a constraint is said to be active when it evaluates to its boundary value. The expert system passes this information on to the designer, and also uses it to suggest approaches for dealing with conflicting specifications or improving the design. A typical message generated by the expert system to describe a conflict in the specifications, following an analysis of the Lagrange multipliers (from the quadratic programming algorithm QPSOL), is shown below :

```
The frequency domain constraints on DY[2,1] could not
be satisfied as they conflict with :
    the frequency domain constraints on DY[1,1]
    the time domain constraints on NU[1,1]
Of these, the Lagrange multipliers hint that the
frequency domain constraints on DY[1,1] are the most
probable cause of the conflict.
It is unlikely that the asymptotic constraints are
responsible for the conflict.
```

4.5. Formulating the Design Specifications.

To assist the user in composing and refining a specification, the expert system considers a list of common design attributes, the characteristics of the plant, and the current specification. These are implemented as a collection of frames [56]. The expert system does not have knowledge of the physical requirements of the control system, such as the limitations of the actuator. For example, while the expert system knows that a good design will often constrain the magnitude of the control signal response following a disturbance, and that to improve the speed of response it may be necessary to relax this constraint, it does not know whether a particular value for the constraint is satisfactory or not. Thus it can only remind the designer of these issues; it is then up to the designer to take the appropriate engineering decisions.

Two mechanisms are employed to assist the designer in formulating a specification. These are invoked by the NEXT STEP and SUGGEST commands, and are discussed in detail below. In addition, the SUGGEST command also provides advice on dealing with

Chapter 4. The Expert System.

conflicts in the specification, and the COMPLETE command a means for checking that the important design aspects have been covered.

4.5.1. The NEXT STEP Command.

A step-by-step design mode has been implemented to help novice designers produce a reasonable design, and at the same time assist them in learning to use the design tool effectively. This procedure guides the user through a sequence of design steps; the process is initiated and perpetuated by repeating the NEXT STEP command, which then recommends the next design step to be taken, explaining why it may be necessary. The designer is free and encouraged to experiment with other commands during this process. Usually the recommended design step is not executed automatically, but the designer is simply shown how to perform it, and is then expected to follow the instructions if required. In this way it is hoped that the user will learn to use the design tool independently.

An example of the use of the NEXT STEP command is shown below :

```
DY[1,1](frequency) >NEXT STEP
Do you want the amplitude of the control signal
to be limited following a step input ? /EXPLAIN?

Constraints on the RU and NU step responses limit the
control action taken following a command input change.
This in turn slows down the closed loop system.  Since
real actuators cannot provide arbitrarily large control
actions, it is good practice to constrain this
response; then (hopefully) the design will be less
susceptible to saturation problems when implemented.

Do you want the amplitude of the control signal to be
limited following a step input ? YES

You need some specifications which will limit the
control signal at output U following a step input at R.
The best way to do this is to constrain the time
response of the RU signals; usually it is necessary to
place constraints on only the initial part of this
response. Another way to achieve this is to constrain
the frequency response, in particular the high
frequency portion. The sequence of commands to use is
     >RU
     >TIME
     >EDIT ALL
```

In this example, the suggested sequence of commands first selects the RU step (time domain) response. The EDIT command is then used to enter or modify the constraints on this response, which are presented in a spreadsheet format by the CACSD package. The qualifier ALL indicates that the constraints set will apply to all elements RU[i,j] of the response.

Further examples of the NEXT STEP command may be found in chapter 6, where it is used extensively in the sample design session listed there.

4.5.2. The SUGGEST Command.

Similar to the step-by-step design mode, the SUGGEST command provides advice to the user on how to improve the design. Guided by appropriate questions, and considering the state of the design in terms of the current specifications and performance, the system can offer suggestions on how to deal with conflicting constraints, which specifications to modify in order to improve performance, or how to meet specifications not explicitly covered by the CACSD package. For example,

```
RY[1,1](time) >SUGGEST
Checking plant DC gain.
A steady state value of at least 2 will be required at
some input of the plant to achieve asymptotic tracking
following a unit step input.
Is this value acceptable ? YES
Do you require asymptotic rejection of disturbances at
input D ? YES

Complete asymptotic disturbance rejection
specifications have not yet been set. These may be
specified using the SET ASYM REJECTION command.

....

DY[1,1](frequency) >SUGGEST
Do you wish to improve the closed loop performance of
the design ? YES
The design is completed in two independent sections,
referred to as Q1 and Q2. These correspond to the
closed loop responses as follows :
    Q1 : responses RY and RU
    Q2 : responses DY, NY, NU, VY, and VU
Which of these sections (1 or 2) are you interested in
? 2
Do you want me to check the design for unusual
combinations of optimization specifications ? YES

I did not find any abnormalities in the optimization
specifications.

Which column of the closed loop response are you
interested in ? 2
Are you prepared to relax any of the constraints in
order to improve the design performance ? /EXPLAIN
The analysis of the Lagrange multipliers from the last
SOLVE command indicates that some of the constraints
are limiting the design performance.

Are you prepared to relax any of the constraints in
order to improve the design performance ? YES
It should be possible to improve the disturbance
rejection performance by relaxing the time domain
constraints on column 2 of the NU responses.
Would you like to try this ? YES
```

Chapter 4. The Expert System. 47

> Use the EDIT command to relax the time domain
> constraints on NU[1,2].

In most cases a particular suggestion is made only once; the designer is expected to take note of it at that time, and keep it in mind during the remainder of the design process. The decision to do this was taken in view of those situations where the designer does not wish to follow the recommendation given, and should not be repeatedly reminded of it. This should suit experienced designers, but not the novice designer who inadvertently ignores important advice. Some relief for this latter case is available, however; after all the relevant suggestions have been made, the designer is given the option of cycling through these suggestions once more.

There are a further two more specific SUGGEST commands. The SUGGEST OPT command provides advice on using the optimization facilities to achieve the required performance. The SUGGEST EDIT command provides advice on the possibilities for changing the performance by placing constraints on the selected response. Examples of these commands are shown below.

> NY[1,1](time) >**SUGGEST OPT**
> I assume that you want to optimize the command
> tracking, disturbance rejection, and decoupling
> properties of the closed loop system.
>
> I suggest that you use the following sequence of
> commands to set optimization for the Q1 part of the
> design.
> >　　　　>RY
> >　　　　>TIME
> >　　　　>SET OPT 1 ALL
> >　　　　>SET OPT 0.1 DIAG
> This sets an optimization weight of 0.1 for the
> diagonal elements, and a weight of 1 for the others.
> You may need to refine this ratio of weights to
> tradeoff the tracking response versus the cross-
> coupling performance. Increasing the weight of the
> diagonal elements will improve the tracking performance
> at the expense of worse decoupling, and vice-versa.
>
> ....
>
> DY[1,1](frequency) >**SUGGEST EDIT**
> The DY response relates to the closed loop system's
> ability to reject disturbances at input D. The designer
> should ensure that low frequency disturbances are
> adequately suppressed; -20 dB is a typical requirement
> for this range. At mid frequencies the designer must
> ensure that the responses do not peak excessively;
> anything over +3 dB is cause for concern. Robustness
> constraints on the singular values may also be set; use
> the EDIT SVD command for this.
>
> In general the ability to reject high frequency
> disturbances is limited. The gain of the diagonal
> elements should be no more than about +3 dB over this

range; this can also be reduced by constraining the
high frequency response of the diagonal elements of NY.
The responses of the off-diagonal elements will depend
strongly on the high frequency behavior of the plant,
but should generally be less than 0 dB. A good way to
get reasonable behavior over this high frequency range
is by limiting the gain of the high frequency portion
of the NU response.

Note that the NY and DY frequency responses are closely
related by the equation
$$NY = I - DY$$
where I is the identity matrix; constraints on the off-
diagonal elements of these responses are therefore
completely equivalent.

4.5.3. The COMPLETE Command.

The expert system is able to check that the design is complete, or at least that it is not incomplete. This is done by analyzing the specifications given to see that all the important design aspects have been covered. The designer is warned of possible omissions in the design specification, and advice is provided on how the design may be updated. In some cases the designer is simply required to check that a particular response is satisfactory.

The checking facility is initiated using the COMPLETE command; it may also be accessed via the HELP system. Some typical dialogue initiated by the COMPLETE command is shown below.

```
NY[1,1](frequency) >COMPLETE
Do you require a robust design ? YES

To ensure some robustness of the design, the SVD plot
of the NY frequency response should not show any
excessive peaking, and the high frequency gain should
taper off. Robustness specifications can be entered as
constraints on the maximum singular values of the NY
frequency response. Details of this approach can be
found in the paper by J.C.Doyle and G.Stein
"Multivariable Feedback Design : Concepts for a
Classical/Modern Synthesis", IEEE Trans. AC, vol. 26,
no. 1, Feb 1981, pp. 4-16. The sequence of commands to
do this is
        >NY
        >FREQ
        >EDIT SVD
The singular values of a frequency response may be
plotted using the SVD command. Similar constraints on
the singular values of the DY response can also be
used.
```

[Here the user is referred to a well-known text [57] on robust design.]

```
. . . .

NY[1,1](frequency)  >COMPLETE
```

Chapter 4. The Expert System.   49

>   The time domain response of RY will be displayed now.
>   Answer YES to the question following if you are happy
>   with the response.
>
> *[At this stage the RY step response is displayed.]*
>
>   Were you satisfied with the response shown ? **YES**
>   ....

4.6. Expanding the Scope of the CACSD Package.

The expert system expands the scope of the CACSD package through checking on aspects of the design not explicitly covered by the specification, and effectively extending the range of specifications which may be used.

In every control system design method, some specifications are treated explicitly, and others implicitly. The latter can only be satisfied by judicious choice of the former, making design somewhat of an art. For example, the CACSD package used deals with constraints on individual closed loop responses directly; to meet specifications on the singular values of the open or closed loop frequency responses requires careful design of the individual closed loop responses.

There may also be some specifications which a given design method cannot address; for example the CACSD method used here cannot deal with requirements on the controller structure. A higher level expert system could be used in turn to select the most appropriate design method in view of the characteristics of the plant and design specifications.

The expert system does provide advice on how to meet some of the implicit constraints, and checks that the design actually satisfies them. For example, the designer may wish to impose constraints on the singular values of the open loop frequency response. These implicit constraints are treated indirectly in the sense that they are not considered by the CACSD package when solving for a controller. However, when set, the expert system can provide advice on how the individual closed loop responses should be constrained in order to meet them. In addition, after a new controller has been computed, the expert system checks that they were satisfied. If not, advice on possible modifications to the specification to account for them is also available if desired. The example below shows the expert system checking on the constraints set on the singular values (SVD) of the open loop (GK) frequency response :

>   DY[1,1](frequency) >**SUGGEST**
>   The SVD constraints on GK have not been satisfied. Do
>   you want help with these ? **YES**
>   Is the problem over the low or high frequency part of
>   the SVD plot ? (enter /EXPLAIN to see the graph) **LOW**
>
>   Constraints (minimum of the minimum singular value) on
>   the low frequency part of the GK response can be met by
>   placing constraints on the low frequency part of the DY
>   response. In general, it is necessary to improve the
>   performance of the Q2 section of the design; this goal

```
may also be achieved through optimization of the DY or
NY step responses.
```

4.7. Optimizing the Use of the CACSD Subroutines.

In general terms, the commands presented to the expert system are translated into a sequence of calls to the CACSD package. In some instances, intermediate results from previous computations may be available and convenient to use in executing the current command, thus improving the efficiency of the design package. The expert system contains knowledge of these situations, and is programmed to exploit them.

An example of this is in the computation of the controller using the SOLVE command. Firstly, only those elements of the **Q** matrix relating to specifications which have changed since the previous controller computation require re-evaluation. Secondly, when sufficient memory is available, it is possible to store the constraints and optimization functions generated when the specifications are translated into the Q domain. Then, only those specifications which have changed since the previous controller computation need re-translation.

The expert system is also useful in dealing with exception conditions in the numerical software; for example, there may be insufficient memory available when the design tool is used on large problems, or the SOLVE operation may have been interrupted prematurely by the user. The expert system is able to identify these situations, and advise the user on how best to proceed.

4.8. Summary.

The most important functions of the expert system have been outlined in this chapter. The expert system forms an interface to the CACSD package, and assists both novice and experienced designers in using the design method. Based upon a database of common design features, the NEXT STEP and SUGGEST commands are able to guide and assist the user in formulating and refining the design specification, and in dealing with conflicting performance constraints. Similarly the COMPLETE command helps the user to check that the design is complete. The expert system has also been used to effectively extend the scope of the design method, as well as to integrate information from various analyses of the design.

CHAPTER 5
IMPLEMENTATION OF THE EXPERT SYSTEM

5.1. Introduction.

This chapter discusses the implementation of the expert system using the facilities provided by the CXS expert system shell. It begins by listing the important features to be considered when selecting an expert system shell, and then describes two of those features - the communication with external programs, and the database facilities - in detail. A discussion of the overall structure of the expert system follows, which includes examples of typical rules form the knowledge base. Here particular emphasis is placed on those sections of the expert system which assist the designer in formulating the design specification, and in dealing with any conflicts in the performance constraints.

5.2. Selection of the Expert System Shell.

The following properties were considering important in selecting an expert system shell for implementing an intelligent design system :

- The shell should provide an efficient and effective method of communication with external, preferably memory-resident, programs. Numerical analysis algorithms are best implemented in a conventional programming language, and the expert system should be able to interact with these external programs easily.

- The shell must be suitable for implementing non-monotonic logic systems. Being a feedback process [29], control system design forms an inherently non-monotonic logic system [19].

- The shell should provide flexible knowledge representation capabilities. The nature of the multivariable problem makes database facilities attractive.

- The shell should have flexible user interface facilities. A significant part of the expert system is devoted to interaction with the designer, and this portion should be both easy to use and program.

- The shell must support a large knowledge base, and should execute efficiently. Control system design is a complex procedure, and thus requires a considerable knowledge base.

- The memory requirements (RAM) of the expert system should be modest. Ideally it should be co-resident with the CACSD package in RAM.

- The shell should support at least simple arithmetic computations.

- The shell must run under the MS-DOS operating system on an IBM-PC type computer, and must be very modestly priced. These two constraints were demanded by limited research funding.

None of the existing expert system shells examined for the IBM-PC (VP-Expert, Synapse, K-Shell, dmX, and Personal Consultant) satisfied all of these requirements. A solution was eventually obtained by extending the CXS expert system shell written previously by the author [58]. Some of the special features of this shell, and how they are used in building the expert system, are detailed below.

CXS is a backward-chaining rule-based expert system shell for the MS-DOS operating system. At present the knowledge-base contains approximately 400 rules.

5.3. Communication with External Programs.

An efficient interface to external software is essential for a high-performance design system. CXS allows linking to external memory resident programs, in this case the CACSD package described in chapter 3, through a remote procedure call scheme based on MS-DOS interrupts. The external programs have access to an array of expert system variables, with elements named T0, T1, ... T99, each of which may contain symbolic or numeric values. This interface is used by the expert system to execute functions such as loading and saving transfer functions, plotting step or frequency responses, translating the specification into a quadratic programming problem, or examining the status of a performance constraint. Details of this interface are given in appendix B. A typical set of rules using this interface, in this case to plot the singular value (SVD) frequency response, is shown below.

```
     IF LINE cmdline ["svd"]      ;If the SVD command was entered
     THEN
        FIND ^set_response        ;Use the rule below to assign the currently
                                  ;selected response to the array variables.
        T0 = "freq"               ;Always use the frequency response
        INTR 96,35,0              ;Call interrupt 96, function 35 to select the
                                  ;response
        INTR 96,1021,0            ;Call interrupt 96, function 1021 to plot the
                                  ;SVD response
        cmd_action is "done"
     END
```

Chapter 5. Implementation Of The Expert System. 53

```
CALC                              ;CALC = IF with no conditions
   T0 = domain
   T1 = resp                      ;Assign the currently
   T2 = i_num                     ;selected response to
   T3 = o_num                     ;the array variables T0-T3
   set_response is "done"
END
```

The ^ operator used in the FIND ^set_response statement above signifies that a new value for the variable set_response must be computed. This operation, similar to the RESET statement, is used frequently in this non-monotonic logic system.

5.4. Database Facilities.

The expert system stores knowledge in rules and facts; in addition, facts may be collected into databases, using the Prolog-style database facilities which CXS provides. Five databases are used within the expert system, and these are examined below.

5.4.1. The RESP_DB Database.
This database holds the symbols corresponding to each closed loop response, and the domain, on which performance constraints may be placed. These are the RY, RU, DY, NY, and NU responses; each of these appears separately for the time and frequency domains.

This database is used internally by the expert system, mainly to access sets of responses using a common rule. The fields in the RESP_DB database are listed below.

Table 5.1. Fields in the RESP_DB database.

Field	Description
DOMAIN	"Time" or "frequency".
RESPONSE	The particular closed loop response, for example "DY".

5.4.2. The SPEC_DB Database.
Much of the design specification is stored in this database; the constraints on the responses are stored within the CACSD package. The records of the SPEC_DB database contain the following fields :

Table 5.2. Fields in the SPEC_DB database.

Field	Description
DOMAIN	"Time" or "frequency".
RESPONSE	The particular closed loop response, for example "DY".
COLUMN	The input number, 1...n
ROW	The output number, 1...n
Q	The Q matrix corresponding to this response, i.e. 1 for **Q1** or 2 for **Q2**.
STATUS	The status of the constraints on this response : "satisfied", "active", "unsatisfied", or "unknown".
ASYM	The asymptotic value required for this response. If none, then the value "none" is stored.
OPT	The optimization weight for this response. If none, then the value "none" is stored.
RELAX	If the designer has indicated that constraints on this response cannot be relaxed, the value "no" is stored; otherwise "unknown".

The CXS shell provides pattern matching facilities which are useful for analyzing the specification. A typical rule from the knowledge base, using this database and a pattern matching condition, is :

```
IF can_relax_some is "yes"
AND opt_q is 1
AND q1_objective is "tracking"
AND opt_col isnt "unknown"
AND FIND spec_db
      [=T0,"RU",opt_col,*,*,"active",*,*,"unknown"]
AND can_relax_ru isnt "no"
THEN
   sugst_opt is "done"
   DISPLAY
Use the EDIT command to relax the \%T0% domain
constraints on column \%opt_col% of the RU response.
This should improve the tracking performance.
END

ASK can_relax_ru ["yes","no"]
It should be possible to improve the tracking
performance by relaxing the \%T0% domain constraints on
column \%opt_col% of the RU response.
Would you like to try this ?
```

This rule is used to suggest that, if the designer is prepared to relax some constraints, and wants to improve the tracking performance, it may be necessary to relax the constraints on the RU response. Note that if the FIND clause finds a record in the SPEC_DB database with fields

RESP = "RU",
COL = the current value of the variable opt_col,

Chapter 5. Implementation Of The Expert System.

```
        STATUS = "active",
and     RELAX  = "unknown",
```

then it assigns the value of the DOMAIN field is to the variable T0. Also shown above is the corresponding question which is put to the user when searching for a value for the variable `can_relax_ru`. The sequence `\%var_name%` in the text displays the value of the particular expert system variable (T0 and `opt_col` in this example).

The expert system uses rules of the form shown above, together with knowledge of the design constraints and present performance, to analyze the specification and current state of the design. While the actual constraints on the responses are stored within the CACSD package, the expert system has efficient access to these through the remote procedure call facility. Similarly, the performance of the design is analyzed within the CACSD package, and the results then transferred to the expert system. In some cases the designer is shown a response graph, and then asked qualitative questions about it, for example whether or not the high frequency gain is considered excessive for the particular application.

5.4.3. The SOLVE_DB Database.

The status of each column of the **Q1** and **Q2** sections of the design are stored in this database. In particular, and for each of these columns, the database records whether or not the specification has been changed since the last SOLVE command, the result of the last SOLVE command (if it was successful, or the nature of the problem), the particular response which could not be satisfied (if any), and the likelihood of the asymptotic constraints being responsible for the conflict. The fields in the SOLVE_DB database are listed below.

Table 5.3. Fields in the SOLVE_DB database.

Field	Description
Q	The **Q** matrix (**Q1** or **Q2**)
COLUMN	The column of **Q** (1..N)
CHANGED	"YES" if the specifications for this column of **Q** have been changed; else "NO".
FAIL_CODE	A code indicating the reason for failure (including no failure).
ASYM_RATIO	The ratio of the Lagrange multiplier for the asymptotic constraints compared to the average multiplier values.
FAIL_DOMAIN	The domain of the constraint not satisfied (if any).
FAIL_RESP	The response not satisfied (if any).
FAIL_ROW	The row of the constraint not satisfied (if any).

5.4.4. The PARAM_DB Database.

This database records the values of various design parameters; it also contains the function code used to pass the value on to the CACSD package, and an indication of whether or not the value has been changed during the design session. At present the parameters stored are the sample time, the maximum modulus allowed for the closed loop poles, the number of points at which frequency responses are evaluated, the number of samples considered for the step responses, and the QSTEP and NVARS parameters. The PARAM_DB database has the following fields :

Table 5.4. Fields in the PARAM_DB database.

Field	Description
SYMBOL	The symbol name (e.g. QSTEP).
VALUE	The value of the symbol.
INTR_NO	The interrupt function number in the CACSD package to use to set the value.
CHANGED	Whether or not this value has been changed.

5.4.5. The FEATURE_DB Database.

The records in this database are used as frames [56] to capture knowledge relating to various design features, and have the following slots :

Table 5.5. Fields in the FEATURE_DB database.

Field	Description
WANTED	The symbol to use to check whether or not the designer wants this feature. Where this feature is always required, "YES" is stored.
SPECIFIED	The symbol to use to check if the current specifications already account for this feature.
Q	The section of the design (**Q1** or **Q2**) to which this feature corresponds.
SATISFIED	Whether or not the designer is happy with the current performance of the design with respect to this feature.
ADVISED	Whether or not this advice has already been given.
ADVICE_RULE	The rule to use to display the relevant advice.
STEP	The step number for use in the NEXT STEP module. If not to be used there, then "none".
POSSIBLE	The symbol to use to check whether or not this feature is possible. Where this feature will always be possible, "YES" is stored.
RESPONSE	The closed loop response in question.
DOMAIN	The domain (time or frequency) of the response above.
SVD	Whether or not the feature is based on the singular values of the frequency response.

The design features presently represented in this database are listed below. The list is by no means complete; it may be argued that no such list could ever be complete. Unfortunately the strict memory limitations of MS-DOS prevent it from being expanded much further in the present expert system. The FEATURE_DB database is used by the NEXT STEP, SUGGEST, and COMPLETE modules of the expert system. The design features considered are :

Chapter 5. Implementation Of The Expert System.

- *Asymptotic properties.* If possible and desired, the design should include asymptotic tracking of command signals, asymptotic rejection of disturbances, and asymptotic decoupling. It may not always be possible to achieve this on all responses simultaneously, depending on the dc gain of the plant. In this case the designer is advised to set the asymptotic requirements on the individual responses as required.

- *Disturbance rejection properties.* The design should ensure that low frequency disturbances are adequately rejected. To achieve this, the design should include constraints on the low frequency sections of the DY frequency responses.

- *Control signal magnitude.* The design should limit the magnitude of the control action. This is often desirable to reduce actuator saturation problems when implemented, and generally also improves the robustness of the design. Usually the RU and NU step responses are constrained; it is also possible to achieve the same effects by constraining the high frequency sections of these responses.

- *Optimization.* Each column of the design should contain some optimization specification, to ensure that all available degrees of freedom in the design method are used.

- *Robustness.* To give the design some robustness to modeling errors, the singular values of the NY frequency response should not exhibit excessive gain, particularly at high frequencies. The singular value response should be constrained; constraints on the individual elements of the NY frequency response will probably be necessary to achieve this.

- *The DY singular values.* The user should check the singular values of the DY frequency response to ensure that the design has adequate low frequency disturbance rejection. In addition and similar to the feature above, the gain at mid and high frequencies should not be excessive, for robustness considerations.

- *The RY and NY step responses.* If these step responses do not exhibit the desired closed loop performance, they should be constrained as necessary.

- *The NVARS and QSTEP parameters.* The values of these parameters can sometimes have a profound effect on the quality of the design. Large values of NVARS, the number of decision variables in each element of the Q matrices, produce better designs but increase the time required to find the solution. In general, large values of QSTEP, used in the parameterization of the Q matrices, suit slow control systems; the designer should experiment with different values. The user is reminded of their importance if these parameters have not been changed during the design. Otherwise it is assumed that the user has already found suitable values.

5.5. Structure of the Expert System.

Figure 5.1 illustrates the overall structure of the expert system. The command line interpreter constitutes the bulk of the program. This in turn may be divided into four main components: the help module, the suggest and next step modules, the module to explain conflicts in the design specification, and the module to check that the specification is complete. The command line interpreter also deals with commands which are to passed on to the CACSD package.

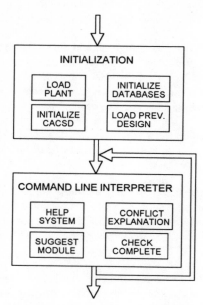

Figure 5.1. Structure of the expert system program.

5.5.1. Initialization.

The initialization process depends to a large extent on whether a completely new design is required, or the design is a continuation of some previous session.

For a new design, the user must enter the plant transfer function, and that of a nominal stabilizing controller if necessary. The expert system checks that the plant is suitable (for example the transfer function must be strictly proper), and initializes its databases to reflect no specifications on any of the closed loop responses.

It is also possible to continue with a previously saved design. Here the relevant transfer functions are retrieved from their files, and databases are loaded with their previous values.

Some initialization is common to both cases. This includes computing the matrix fractions and the nominal step and frequency responses according to the formulae listed in table 3.2, for example. The CACSD package, which is called to perform many of the tasks mentioned above, also requires initialization.

Chapter 5. Implementation Of The Expert System.

5.5.2. The Command Line Processor.

Pattern matching facilities, similar to those for databases, are used to analyze the user command line. For example, the GROUP PLOT command, and its abbreviation GP, are treated by the rule

```
IF LINE cmdline["group","plot"]
OR LINE cmdline["gp"]
THEN
   INTR 96,1006,0          ;Plot the group of responses
   cmd_action is "done"
END
```

The more complex commands are treated in a similar fashion. For example, the OPT command used to enter the optimization specifications, has the syntax

    action OPT [value] [group]

where action is either SET or RESET,
        value is the optimization weight (specified for the SET action only),
and   group is ALL, DIAG, OFFDIAG or COLUMN, if specified.

Again the pattern matching facilities provided by CXS in the form of the LINE condition are invaluable for decoding these commands. In this case, the rule

```
IF LINE cmdline[=T0,"opt"]
AND T0 isnt "set"
AND T0 isnt "reset"
THEN
   ....
END
```

could be used to check for an illegal OPT command.

5.5.3. The HELP Module.

The menu driven online help system, discussed in the previous chapter, is initiated by the HELP [topic] command. A useful feature of the CXS shell is that text for display purposes need not be stored in memory, but can be recalled from disk when required. This makes it possible to provide a large amount of information on a wide range of topics, even with the restrictive MS-DOS memory limitations.

    Further help on any question posed to the user is also available by typing /EXPLAIN in response to that question.

5.5.4. The SUGGEST and NEXT STEP Modules.

The SUGGEST and NEXT STEP modules are programmed similarly. The expert system stores a list of design features which an experienced designer would consider, and compares these against the current specifications and requirements of the user. In some cases, data from an analysis of the plant is also considered. Using this knowledge, the expert system can determine whether or not a particular design feature should be added to the design specification, and advise the designer accordingly. The advice generally includes the set of commands which must be used to accomplish this.

Rules in the SUGGEST module have the generic structure shown below.

```
IF this advice has not previously been given
AND the design feature is possible with the given
      plant
AND the design feature is not accounted for by the
      current specifications
AND the feature is probably required
THEN
  Note that the advice has been given.
  Display the advice.
END
```

The NEXT STEP rules are very similar; the main difference is the `step_state` expert system variable which is used to record the progress of the advice given. The generic structure of these rules is

```
IF the previous steps in the design process have
      been considered
AND the design feature is possible with the given
      plant
AND the design feature is not accounted for by the
      current specifications
AND the feature is probably required
THEN
  Note that this step has been suggested.
  Display the advice.
END
```

The similar structures of the rules in these two modules can be exploited to give efficient coding. This has been achieved using the FEATURE_DB database of design features described in section 5.4.5. Both modules have rules which search through this database to identify design features which the user should consider. For example, the record

```
["limit_u_reqd", "ru_spec_ok", 1, "unknown", "no",
 "rut1", 5, "yes", "ru", "time", "no"]
```

is used to check that specifications have been placed on the control signals following a step response at input R. This in turn relates to the rules given below. A value for the variable `limit_u_reqd` is found by asking the user. The condition clauses of the rule for `ru_spec_ok` check that there is some element of the RU response which has constraints on neither the initial part of the step response, nor on the high frequency response; this information is obtained via the CACSD module. Finally the rule for `rut1` is invoked to display the advice if all the conditions have been satisfied.

```
ASK limit_u_reqd ["yes","no"]
Do you want the amplitude of the control signal to be
limited following a step input at R or D ?
EXPLAIN
```

Chapter 5. Implementation Of The Expert System.

Constraints on the RU and NU step responses limit the control action taken following a command input change. This in turn slows down the closed loop system. Since real actuators cannot provide arbitrarily large control actions, it is good practice to constrain this response; then (hopefully) the design will be less susceptible to saturation problems when implemented.

```
IF FINDSOME spec ["time","RU",=T2,=T3]
AND ^low_t_max_spec isnt number
                      ;no constraints on maximum step response
AND ^low_t_min_spec isnt number
                      ;no constraints on minimum step response
AND ^high_f_spec isnt number
                      ;no constraints on max high frequency response
THEN
   ru_spec_ok is "no"
END

CALC
# rut1
   DISPLAY
You need some specifications which will limit the
control signal at output U following a step input at R.
The best way to do this is to constrain the time
response of the RU signals; usually it is necessary to
place constraints on only the initial part of this
response. Another way to achieve this is to constrain
the frequency response, in particular the high
frequency portion. The sequence of commands to use is
         >RU
         >TIME
         >EDIT ALL
END
```

In addition to the types of advice mentioned above, there are also the SUGGEST OPT and SUGGEST EDIT commands. SUGGEST OPT deals with the optimization of responses. The designer may choose to optimize either the RY response (for tracking performance) or the RU response (for low control power); a similar choice is made between NY (or DY) and NU. Given this choice, the expert system can also check that the optimization specifications do not contradict the choice of objective. Optimization of step responses generally produces better results than optimization of the corresponding frequency responses; the expert system points this out if necessary, and advises (but does not compel) the user accordingly. The command also provides advice on how the performance of the response selected for optimization may be improved further. This involves relaxing those constraints which are limiting the performance, adjusting the relative values of the optimization weights, or experimenting with the NVARS and QSTEP parameters.

A typical rule from this knowledge base, which deals with the absence of any optimization on column q1opt of the **Q1** section of the design, is shown below.

```
    IF q1opt is number              ;No optimization has been specified for
                                    ;column q1opt
    AND q1_objective is "tracking"
    THEN
       sugst_opt is "done"
       DISPLAY
    I suggest that you use the following sequence of
    commands to set optimization for column \%q1opt% of the
    Q1 part of the design.
            >RY
            >TIME
            >\%q1opt% \%q1opt%
            >SET OPT 1 COLUMN
            >SET OPT 0.1
    This sets an optimization weight of 0.1 for the element
    on the diagonal, and a weight of 1 for the others. You
    may need to refine this ratio of weights to tradeoff
    tracking response versus cross-coupling performance.
    Increasing the weight of the diagonal elements will
    improve the tracking performance at the expense of
    worse decoupling, and vice-versa.
    END
```

SUGGEST EDIT gives information about the currently selected response, indicating its relevance to the final design, and discussing why the designer may wish to consider placing constraints on it. A typical example of this rule, dealing with the case when the DY frequency response is currently selected, is

```
    IF resp is "DY"
    AND domain is "freq"
    THEN
       sugst_edit is "done"
       DISPLAY
    The DY response relates to the closed loop system's
    ability to reject disturbances at input D. The designer
    should ensure that low frequency disturbances are
    adequately suppressed; -20 dB is a typical requirement
    for this range. At mid frequencies the designer must
    ensure that the responses do not peak excessively;
    anything over +3 dB is cause for concern. Robustness
    constraints on the singular values may also be set; use
    the EDIT SVD command for this.

    In general the ability to reject high frequency
    disturbances is limited. The gain of the diagonal
    elements should be no more than about +3 dB over this
    range; this can also be reduced by constraining the
    high frequency response of the diagonal elements of NY.
    The responses of the off-diagonal elements will depend
    strongly on the high frequency behavior of the plant,
    but should generally be less than 0 dB. A good way to
    get reasonable behavior of this high frequency range is
    by limiting the gain over the high frequency portion of
    the NU response.
```

Chapter 5. Implementation Of The Expert System.

```
Note that the NY and DY step responses are closely
related by the equation
      NY = I - DY
where I is the identity matrix; constraints on the off-
diagonal elements of these responses are therefore
completely equivalent.
END
```

5.5.5. The Module Checking for Completeness.

The purpose of this module is to check that the design is complete in the sense that a list of design features have been considered. The structure of rules in this module, as shown below, is similar in many ways to those of the SUGGEST and NEXT STEP modules. Here again the design specifications are compared against a list of common design features in the FEATURE_DB database, and the user advised accordingly. The rules in this module have the generic form shown below.

```
IF the design feature is not accounted for by the
      current specifications
AND the design feature is possible with the
      given plant
AND the feature is probably required
AND the designer is not happy with the current
      performance
THEN
   Display advice on how to update the specification.
END
```

5.5.6. The Module Explaining Specification Conflicts.

Conflicts in the specification are detected at two levels by the CACSD package, in both cases as conflicts in the set of linear constraints generated for the particular design. The expert system has the task of explaining these conflicts, and of suggesting ways for dealing with them. The expert system also deals with other complications which preclude a solution to the design; for example there is the possibility of insufficient RAM memory being available for large design problems. In this instance the user is advised on ways to reduce the memory requirements.

Some conflicts are detected when the design specification is translated into linear constraints. These conflicts relate to linear constraints in the form

$$\mathbf{a}^T \mathbf{x} \leq \alpha < 0,$$

with $\mathbf{a} = \mathbf{0}$,

which cannot be satisfied by any vector $\mathbf{x}$. In the time domain this situation typically arises with a rise time specification faster than the plant dead time. In the frequency domain this type of conflict domain results from poles or zeros of the plant on the unit circle fixing the system response at those frequencies. Similar conflicts also arise occasionally in the asymptotic specifications.

When the CACSD package detects this type of conflict, the expert system is notified of the response concerned. The messages displayed to the user, for the time and frequency domain cases respectively, are shown below. In each case the designer must relax or remove the offending specification.

> It is not possible to satisfy the specifications on the RY[1,1] time domain response, due to the characteristics of the plant (usually the plant dead time). You will have to relax the specifications on this response.
>
> It is not possible to satisfy the specifications on the DY[2,1] frequency domain response, due to the characteristics of the plant (generally a pole or zero of the plant on the unit circle). You will have to relax the specifications on this response. It may be possible avoid this problem by placing the specifications over ranges which exclude this pole or zero frequency.

Most conflicts are detected when QPSOL, the quadratic programming algorithm, finds no feasible solution to the quadratic programming problem. To facilitate analysis of these conflicts, the linear constraints are grouped according to the design specifications that they represent. Thus all constraints relating to specifications on the RY[2,2] step response will be in one group, and all those on the NU[1,2] frequency response will be in another. QPSOL tackles phase I of the quadratic programming problem by finding a vector **x** which satisfies all constraints in the first group, then one which satisfies the constraints in the first two groups, and so on, until all groups have been considered, or a conflict has been found.

The expert system has access to the status of each of these groups following a call to QPSOL; this information is requested from the CACSD package and stored in the SPEC_DB database. The linear constraints are given a status of active, satisfied, or not satisfied, according to the terminology of the active set method used. Consider the constraint

$$c(\mathbf{x}) \leq \alpha.$$

This constraint, when evaluated at $\mathbf{x}_1$, is assigned a status according to table 5.6.

Table 5.6. Status of constraint $c(x_1)$.

$c(\mathbf{x}_1)$	Status
$= \alpha$	active
$\leq \alpha$	satisfied
$> \alpha$	not satisfied

Note that only those constraints appearing in the active set (equation 3.34) are considered active, and that the active status implies that it is also satisfied. Thus it is possible that some constraints may be met with equality, but not be considered active. The status of a group is determined from the status of the corresponding linear constraints, using the following rules :

- If the group has not yet been considered by QPSOL, then its status is unknown. Else,

Chapter 5. Implementation Of The Expert System.

- If any linear constraints are not satisfied, then the group is not satisfied. Else,
- If any linear constraints are active, then the group is active. Else,
- The status is satisfied.

The expert system uses this data to inform the designer of the conflict. A typical message, generated by the expert system to describe this conflict, is shown below :

```
The frequency domain constraints on DY[2,1] could not
be satisfied as they conflict with :
    the frequency domain constraints on DY[1,1]
    the time domain constraints on NU[1,1]
Of these, the Lagrange multipliers hint that the
frequency domain constraints on DY[1,1] are the most
probable cause of the conflict.
It is unlikely that the asymptotic constraints are
responsible for the conflict.
```

The likelihood of the asymptotic constraints contributing to the conflict, as well as the specification most probably causing the conflict, is determined by analyzing the Lagrange multipliers computed in the QPSOL algorithm. Although subject to scaling, the more positive the value of the multiplier, the more likely that constraint is a limiting factor in the design. The inverse argument is commonly used active set methods, where the constraint with the most negative multiplier is assumed to have the least impact on the solution, and is deleted from the active set.

In addition to explicitly describing the cause of the conflict as shown above, the expert system can also provide the user with some assistance in resolving the conflict. One way this is done is by diagnosing some particular conflicts. For example, given the relationship between the NY and DY responses, specifications placed on both responses may conflict. This is dealt with by rules of the form shown below; the example given detects a specification of less than unity gain at low frequencies on a diagonal element of NY, when asymptotic disturbance rejection has been specified on the corresponding element of DY.

```
IF spec.resp is "NY"
AND spec.col is spec.row
AND spec.domain is "freq"
AND FIND spec_db ["freq","DY",spec.col,spec.row,
                                *,*,number]
AND ^ny_dy_asym_clash is "yes"
THEN
   sugst_conflict is "done"
   DISPLAY
You must change the specifications on the frequency
response of NY[\%spec.row%,\%spec.col%], or remove the
asymptotic rejection specification on
DY[\%spec.row%,\%spec.col%], to make them consistent.
Of these, the constraints on the NY response seem
strange to me.
END

CALC
  T0 = "freq"
  T1 = "NY"
  T2 = spec.col
```

```
        T3 = spec.row
        INTR 96,35,0          ;select NY[i,i](freq)
        T5 = 0
        T6 = 0.004
        INTR 96,2025,1        ;get the minimum value of the constraints at
                              ;low frequencies
        IF T8 < 1
          ny_dy_asym_clash is "yes"
        END
    END
```

Sometimes it is possible that a better choice of the NVARS and/or QSTEP design parameters could resolve the conflict. The CACSD method uses these parameters in the parameterization of the Q matrices, which can influence the design significantly. In this case the designer is advised on how the parameters may be changed.

Further advice is also available on resolving conflicts in the specifications on the singular values of the frequency responses. These specifications are not translated into equivalent linear constraints, but the expert system can provide advice on how to modify the other closed loop specifications in order to meet them. For example, the rule below treats conflicts in specifications on the high frequency region of the open loop (GK) frequency response.

```
        IF svd_resp_c is "GK"
        AND gk_svd_prob is "high"     ;the problem is with the high
                                      ;frequency part of the response
        THEN
          sugst_svd is "done"
          DISPLAY
        Constraints (maximum of the maximum singular value) on
        the high frequency part of the GK response can be met
        by placing constraints on the high frequency part of
        the NY response. In general, it is necessary to slow
        the performance of the Q2 section of the design, thus
        making it more robust. Another way to achieve this is
        by placing constraints on the high frequency part of
        the NU response.
        END
```

5.6. Summary.

The issues involved in implementing the expert system have been described in this chapter. These include the selection of an appropriate expert system shell, communication with the CACSD package, and the use of databases to store common design features and the status of the design. The overall structure of the expert system has been presented, as well as details of the rules used in the NEXT STEP, SUGGEST, and COMPLETE modules. It has also been shown how the active set and Lagrange multipliers of the quadratic programming problem may be analyzed to identify the cause of conflicts within the design specification, and to assist the user in dealing with these.

CHAPTER 6
SAMPLE DESIGN SESSIONS

6.1. Introduction.

This chapter presents two examples illustrating the use of the design system. In the first application a detailed transcript of the interaction between the designer and the expert system is provided. The second example describes the application of the package to an unstable plant.

6.2. Example 1 : A Mine Milling Plant.

To illustrate the use of the expert system, the design system has been applied to a model of a gold mine milling unit [59]. The z domain transfer function of the plant, with a normalized sampling rate of 1 Hz, is given below.

$$G(z) = \begin{bmatrix} \dfrac{0.6512}{z-0.9031} & \dfrac{1.19}{z-0.05306} & \dfrac{0.0503}{z-0.7304} \\ \dfrac{7.037}{z-0.913} & \dfrac{-3.296}{z-0.08696} & \dfrac{0.2055}{z-0.7595} \\ \dfrac{-0.8206z^{-5}}{z-0.8436} & \dfrac{(0.5127z-0.3876)z^{-1}}{z^2-0.5837z+0.0742} & \dfrac{0.2923z-0.264}{z^2-0.6603z+0.1052} \end{bmatrix}$$

The control system design for this plant is carried out using the expert system for guidance; the transcript from the design session is given below. Commands and answers entered by the user are shown in **BOLD** type. Comments added to this transcript are printed in *italics* and enclosed in square brackets.

Welcome to MV-CXS, the Expert System design tool for linear
multivariable control system design.

The file MILL.GZ has been found. Do you wish to load the
plant transfer function ? **YES**
The plant transfer function has been loaded from file
MILL.GZ
Do you wish to modify the plant transfer function ? **NO**
   [For the purposes of demonstration this design will be started from scratch.
   Therefore the two questions following are answered with NO.]
Do you want to load the previous design ? **NO**
The file MILL.PRM has been found. Do you want to load the
previous design parameters from it ? **NO**
Initializing the databases..........
Checking that individual elements of the plant transfer
matrix are coprime and strictly proper... Ok
What is the sampling time (seconds) ? **1**
The plant poles have a maximum modulus of 0.913. Is this
maximum modulus acceptable for the closed loop poles ? **YES**
Would you like help in setting up a specification ? **YES**

The following block diagram illustrates the type of system
under consideration. You will need to remember the meanings
of the inputs labeled R, N, D and V, and the outputs Y and
U.

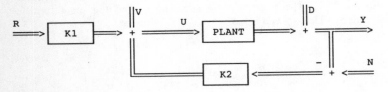

For more help with the design system, type 'HELP'. For help
on selecting a response to work on, type 'HELP SELECT'.
Remember that you may also type '/EXPLAIN' in response to a
question if you need assistance.

Finally, to help you further the command 'SUGGEST' can often
offer advice on what to do next, and the 'NEXT STEP' command
can help you get started with a design.

NY[1,1](time) >**NEXT STEP**

The step-by-step design mode has been selected. This will
give you recommendations on what you should do next; it is
then up to you to enter the commands if you wish. Note that
any of the other commands can also be used; for example you
may examine the various responses using the 'PLOT' command.
Once you have completed the step, the 'NEXT STEP' ('NS' for
short) command should be used once more to recommend the
following step to be taken.

Chapter 6. Sample Design Sessions.

Checking plant DC gain. A steady state value of at least 8.369168 will be required at some input to the plant to achieve asymptotic tracking and decoupling following a unit step input. Is this value acceptable ? **YES**

Would you like asymptotic rejection of disturbances at input D ? **/EXPLAIN**
Asymptotic rejection of disturbances at input D means that, at DC, the frequency response matrix Hdy is zero. Thus the disturbances will be rejected completely, and there will be no steady state cross-coupling (interaction). Usually this is desired; unfortunately it sometimes requires prohibitively large control signal inputs to the plant. If you think you might like asymptotic rejection, answer YES.

Would you like asymptotic rejection of disturbances at input D ? **YES**

It is possible to achieve asymptotic rejection of disturbances at input D.

You can enter asymptotic disturbance rejection requirements using the 'SET ASYM REJECTION' command. If desired, this specification may be lifted later using the 'RESET ASYM REJECTION' command. For more information on the asymptotic specifications, type 'HELP ASYM'.

DY[1,1](frequency) >**SET ASYM REJECTION**
DY[1,1](frequency) >**NEXT STEP**
Would you like asymptotic tracking of the command signal R ?
YES

You can enter asymptotic tracking/decoupling requirements using the 'SET ASYM TRACKING' command. If necessary, this specifications may be lifted later using the 'RESET ASYM TRACKING'. For more information on the asymptotic specifications, type 'HELP ASYM'.

RY[1,1](frequency) >**SET ASYM TRACKING**
RY[1,1](frequency) >**NEXT STEP**
Do you require rejection of low frequency disturbances at input D ? **YES**

Rejection of low frequency disturbances has not been completely specified. You should use the 'EDIT' command to enter constraints on the low frequency response of the DY signals to ensure reasonable disturbance rejection properties for the closed loop system. The sequence of commands to use is
>DY
>FREQ
>EDIT ALL
A starting constraint could be -20 dB over the range 0 to 0.0025 Hz.

DY[1,1](frequency) >**EDIT ALL**

[Using the spreadsheet style constraint editor shown below, the maximum value of this response is constrained to no more than -20 dB over the range 0 to 0.0025 Hz.]

```
Frequency domain constraints : DY[1,1]   max = 0.5 Hz
Maximum

| Value (db) | From (Hz) | To (Hz) |
| -20        | 0         | 0.0025  |

Select using [Ctrl] cursor keys : ESC to exit
```

DY[1,1](frequency) >**NEXT STEP**
Do you want the amplitude of the control signal to be limited following a step input at R or D ? **YES**

You need some specifications which will limit the control signal at output U following a step input at R. The best way to do this is to constrain the step response of the RU signals; usually it is necessary to place constraints on only the initial part of this response. Another way to achieve this is to constrain the frequency response, in particular the high frequency portion. The sequence of commands to use is
>RU
>TIME
>EDIT ALL

RU[1,1](time) >**EDIT ALL**
   [Using the constraint editor, the value of the control signal at U is constrained to lie between -10 and 10 units over the range 0 to 50 seconds, following a unit step at input R.]

```
Time domain constraints : RU[1,1]   max = 50 Seconds
Maximum                              Minimum

| Value | From (s) | To (s) |   | Value | From (s) | To (s) |
| 10    | 0        | 50     |   | -10   | 0        | 50     |

Select using [Ctrl] cursor keys : ESC to exit
```

Chapter 6. Sample Design Sessions.

```
RU[1,1](time) >NEXT STEP
```

You need some specifications which will limit the control
signal at output U following a step input at N. The best way
to do this is to constrain the time response of the NU
signals; usually it is necessary to place constraints on
only the initial part of this response. Another way to
achieve this is to constrain the frequency response, in
particular the high frequency portion. The sequence of
commands to use is
```
          >NU
          >TIME
          >EDIT ALL
```

```
NU[1,1](time) >EDIT ALL
```
[As for RU above, the value of the control signal at U is constrained to lie between -10 and 10 units over the range 0 to 50 seconds, following a unit step at input N.]

```
NU[1,1](time) >NEXT STEP
```

You should now use the 'SOLVE' command to take the changes
to the specification into account.

```
NU[1,1](time) >SOLVE
Solving for column 1 of Q1.
Solving for column 2 of Q1.
Solving for column 3 of Q1.
Solving for column 1 of Q2.
Solving for column 2 of Q2.
Solving for column 3 of Q2.
```

```
NU[1,1](time) >NEXT STEP
```

The optimization specifications are not yet complete. Use
the 'OPT' commands to enter your optimization requirements.
For more information on these, type 'HELP OPT'. The 'SUGGEST
OPT' command should be able to help you with this, and will
be started for you now.

I assume that you want to optimize the command tracking,
disturbance rejection, and decoupling properties of the
closed loop system.

I suggest that you use the following sequence of commands to
set optimization for the Q1 part of the design.
```
          >RY
          >TIME
          >SET OPT 1 ALL
          >SET OPT 0.1 DIAG
```
This sets an optimization weight of 0.1 for the diagonal
elements, and a weight of 1 for the others. You may need to
refine this ratio of weights to tradeoff the tracking
response versus the cross-coupling performance. Increasing

the weight of the diagonal elements will improve the
tracking performance at the expense of worse decoupling, and
vice-versa.

```
NU[1,1](time) >RY
RY[1,1](time) >SET OPT 1 ALL
RY[1,1](time) >SET OPT 0.1 DIAG

RY[1,1](time) >NEXT STEP
```

You should now use the 'SOLVE' command to take the changes
to the specification into account.

```
RY[1,1](time) >SOLVE
Solving for column 1 of Q1.
Solving for column 2 of Q1.
Solving for column 3 of Q1.

RY[1,1](time) >NEXT STEP
```

The optimization specifications are not yet complete. Use
the 'OPT' commands to enter your optimization requirements.
For more information on these, type 'HELP OPT'. The 'SUGGEST
OPT' command should be able to help you with this, and will
be started for you now.

I suggest that you use the following sequence of commands to
set optimization for the Q2 part of the design.
```
            >NY
            >TIME
            >SET OPT 1 ALL
            >SET OPT 0.1 DIAG
```
This sets an optimization weight of 0.1 for the diagonal
elements, and a weight of 1 for the others. You may need to
refine this ratio of weights to tradeoff tracking response
versus cross-coupling performance. Increasing the weight of
the diagonal elements will improve the tracking performance
at the expense of worse decoupling, and vice-versa.

```
RY[1,1](time) >NY
NY[1,1](time) >SET OPT 1 ALL
NY[1,1](time) >SET OPT 0.1 DIAG

NY[1,1](time) >SOLVE
Solving for column 1 of Q2.
Solving for column 2 of Q2.
Solving for column 3 of Q2.

NY[1,1](time) >GROUP PLOT
```

Chapter 6. Sample Design Sessions. 73

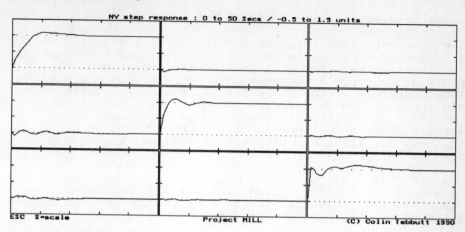

```
NY[1,1](time)   >DY
DY[1,1](time)   >FREQUENCY
DY[1,1](frequency)   >GROUP PLOT
```

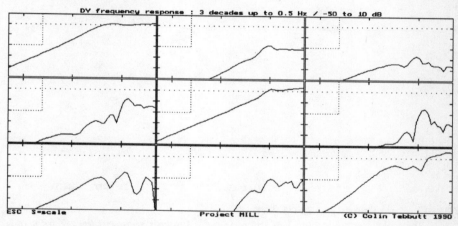

```
DY[1,1](frequency)   >NEXT STEP
```

The 'NEXT STEP' sequence of step is complete. Now you may
wish to look at some of the other performance
specifications; the 'SUGGEST' command could possibly
give you some further advice. If you wish to know more about
tailoring a particular response, select that response and
then type 'SUGGEST EDIT'.

```
DY[1,1](frequency)   >SUGGEST
```
Do you require a robust design ? **/EXPLAIN**
The robustness of a design is a measure of its ability to
handle inaccuracy in the plant model. Since every model of a
real plant contains some inaccuracy, every control system
should have some measure of robustness. So in short, answer
YES!

Do you require a robust design ? **YES**

To ensure some robustness of the design, the SVD plot of the NY frequency response should not show any excessive peaking, and the high frequency gain should taper off. Robustness specifications can be entered as constraints on the maximum singular values of the NY frequency response. Details of this approach can be found in the paper by J.C.Doyle and G.Stein "Multivariable Feedback Design : Concepts for a Classical/Modern Synthesis", IEEE Trans AC, vol 26, no. 1, Feb 1981, pp. 4-16. The sequence of commands to do this is
```
    >NY
    >FREQ
    >EDIT SVD
```
The singular values of a frequency response may be plotted using the 'SVD' command. Similar constraints on the singular values of the DY response can also be used.

NY[1,1](frequency) >**SVD**

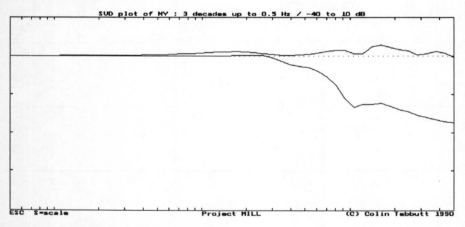

NY[1,1](frequency) >**EDIT SVD**

[The following constraints are entered on the singular values of the NY frequency response :

SVD $\leq +3$ dB, $0 \leq f < 0.1$ Hz,
SVD ≤ -3 dB, $0.1 \leq f < 0.2$ Hz,
SVD ≤ -6 dB, $0.2 \leq f < 0.5$ Hz.]

Would you like advice on how to satisfy these specifications ? **YES**

It is necessary to place/tighten constraints on the individual elements of the NY frequency response in order to meet the SVD constraints on the NY response, over the required range of frequencies. Note that usually the gains of the off-diagonal should be lower than those of the diagonal elements. If there are some elements with relatively high gains, try constraining these first.

Chapter 6. Sample Design Sessions. 75

```
NY[1,1](frequency) >EDIT ALL
```
   [The maximum value of each element in the NY frequency response matrix is constrained to no more than -20 dB over the range 0.1 to 0.5 Hz.]

```
NY[1,1](frequency)  >SOLVE
Solving for column 1 of Q2.
Solving for column 2 of Q2.
Solving for column 3 of Q2.

The frequency domain constraints on DY[3,3] could not be
satisfied as they conflict with :
   the frequency domain constraints on DY[2,3]
   the frequency domain constraints on NY[1,3]
   the frequency domain constraints on NY[2,3]
   the frequency domain constraints on NY[3,3]
Of these, the Lagrange multipliers hint that the frequency
domain constraints on NY[3,3] are the most probable cause of
the conflict. It is unlikely that the asymptotic constraints
are responsible for the conflict.

NY[1,1](frequency)  >EDIT ALL
```
   [The constraints on the NY frequency response are revised as follows :
      $NY(f) \leq -6\ dB,$ $0.1 \leq f < 0.2\ Hz,$
      $NY(f) \leq -12\ dB,$ $0.2 \leq f < 0.5\ Hz.]$

```
NY[1,1](frequency)  >SOLVE
Solving for column 1 of Q2.
Solving for column 2 of Q2.
Solving for column 3 of Q2.
Checking the SVD constraints for section Q2...
All of the SVD constraints have been satisfied.

NY[1,1](frequency)  >SUGGEST
```

The time domain response of RY will be displayed now. Answer YES to the question following if you are happy with the response.

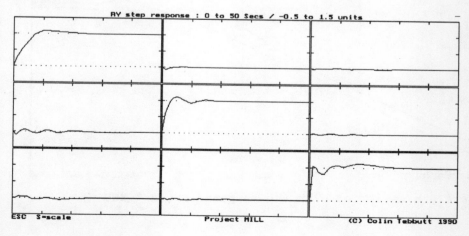

Were you satisfied with the response shown ? **YES**

The time domain response of NY will be displayed now. Answer YES to the question following if you are happy with the response.

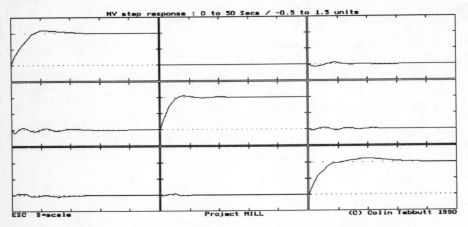

Were you satisfied with the response shown ? **YES**

To ensure good rejection of low frequency disturbances, the low frequency gain of the DY response should be small. In addition, the SVD plot of the DY frequency response should not show any excessive peaking.

The frequency domain response of DY will be displayed now. Answer YES to the question following if you are happy with the response.

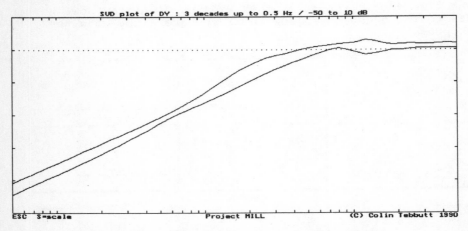

Were you satisfied with the response shown ? **YES**

Chapter 6. Sample Design Sessions.

You can try experimenting with the value of the QSTEP parameter. Larger values for QSTEP suit slower control systems, and smaller values suit fast systems. The present value of QSTEP is 5, and this may be changed using the 'QSTEP n' command. Type 'HELP QSTEP' for more information.

NY[1,1](frequency) >**HELP QSTEP**

The parameter QSTEP is used internally in the numeric software. The usual range is from 1 to about 20; usually the upper limit should satisfy the inequality QSTEP*NVARS <= 100. The larger the value of QSTEP, the slower the control system response tends to be.

For stable plants, the control signal will over the first QSTEP*NVARS sample periods only; thus the combination of QSTEP and NVARS describe a time window over which the control action takes place. Therefore QSTEP and NVARS should be chosen to adequately cover this time window. For unstable plants the interpretation of these parameters is more complex, but the same principles should be applied. Note that the QSTEP and NVARS parameters should be dealt with as a unit; the product QSTEP*NVARS gives the time window, and NVARS the number of subdivisions in this window. For help with NVARS, type 'HELP NVARS'.

The present value of QSTEP may be found by typing 'QSTEP', and the value can be changed using the command 'QSTEP n'. Note that the responses are re-initialized after QSTEP is changed, and so you must use the SOLVE command to find the new controller.

NY[1,1](frequency) >**HELP NVARS**

NVARS is the number of decision variables used in each element of Q in the search algorithm. The higher the value, the better the solution will be, but the longer it will take to find the solution. The maximum value must not exceed 8; the minimum useful value is 2, but generally a value of 5 is reasonable.

For stable plants, the control signal will over the first QSTEP*NVARS sample periods only; thus the combination of QSTEP and NVARS describe a time window over which the control action takes place. Therefore QSTEP and NVARS should be chosen to adequately cover this time window. For unstable plants the interpretation of these parameters is more complex, but the same principles should be applied. Note that the QSTEP and NVARS parameters should be dealt with as a unit; the product QSTEP*NVARS gives the time window, and NVARS the number of subdivisions in this window. For help with QSTEP, type 'HELP QSTEP'.

The present value of NVARS may be found by typing 'NVARS', and the value can be changed using the command 'NVARS n'.

After changing NVARS, you should use the SOLVE command to
find the new controller.

NY[1,1](frequency) >**NVARS 8**
The closed loop responses have been re-initialized. You
should use the SOLVE command to find a new controller.

NY[1,1](frequency) >**QSTEP 4**
The closed loop responses have been re-initialized. You
should use the SOLVE command to find a new controller.

NY[1,1](frequency) >**SOLVE**
Solving for column 1 of Q1.
Solving for column 2 of Q1.
Solving for column 3 of Q1.
Solving for column 1 of Q2.
Solving for column 2 of Q2.
Solving for column 3 of Q2.

Checking the SVD constraints for section Q2...
All of the SVD constraints have been satisfied.

NY[1,1](frequency) >**COMPLETE**

The time domain response of RY will be displayed now. Answer
YES to the question following if you are happy with the
response.

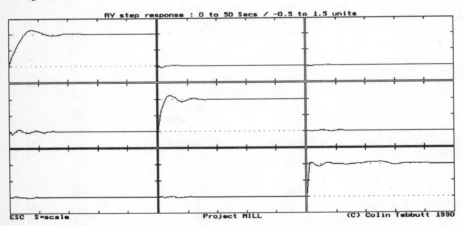

Were you satisfied with the response shown ? **YES**

The time domain response of NY will be displayed now. Answer
YES to the question following if you are happy with the
response.

Chapter 6. Sample Design Sessions.

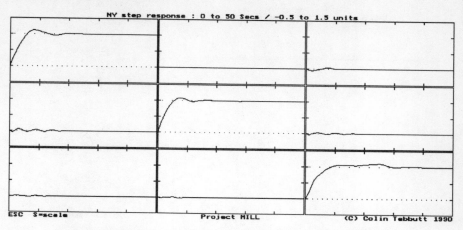

Were you satisfied with the response shown ? **YES**

To ensure good rejection of low frequency disturbances, the low frequency gain of the DY response should be small. In addition, for robustness of the design, the SVD plot of the DY frequency response should not show any excessive peaking.

The frequency domain response of DY will be displayed now. Answer YES to the question following if you are happy with the response.

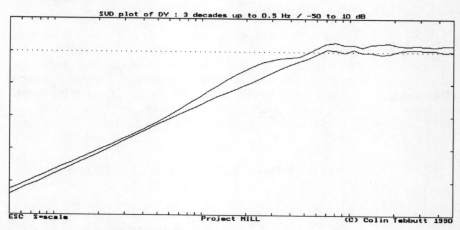

Were you satisfied with the response shown ? **YES**

It seems to me that your design is complete; if you are satisfied with it, use the 'SAVE' command to save it, and then the 'EXIT' command to leave the design session.

NY[1,1](frequency) >**SVD**

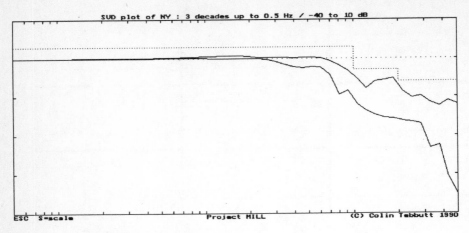

```
NY[1,1](frequency) >SAVE
NY[1,1](frequency) >EXIT
Do you want a report on the current design ? NO
```

[The SAVE command saves the current status of the design for implementation or a possible future design session. The report option, which was not selected, would have produced a printout of the current specification.]

6.3. Example 2 : A Gyroscope.

The 2-gimbal gyroscope studied by Limebeer and Maciejowski [60] is used to illustrate the use of the design system when the plant is not stable. The dialogue with the expert system to engineer a design specification would be quite similar to that given above; it is within the CACSD package that most of the differences between stable and unstable plants are encountered. Hence this section will concentrate on the solution of the control system design problem given a set of performance specifications.

The z domain transfer function of the plant, when sampled at 50 Hz, is

$$\mathbf{G}(z) = \frac{0.001}{d(z)} \begin{bmatrix} 2.4240z^2 - 1.2298z - 0.8155 & 0.7815z^2 + 2.9061z + 0.2076 \\ -0.7835z^2 - 2.9134z - 0.2082 & 0.5170z^2 - 0.2207z - 0.1680 \end{bmatrix}$$

where $d(z) = z^3 - 0.4359z^2 + 0.2384z - 0.8025$. This plant has an unstable pole at $z = 1$, and lightly damped poles at $z = -0.2820 \pm j0.8503$. The nominal controller used to stabilize the plant was

Chapter 6. Sample Design Sessions.

$$\mathbf{K0}(z) = \begin{bmatrix} 0 & 1 \\ -1 & 0 \end{bmatrix} \tag{6.1}$$

Applying the diagonal factorization method gives

$$\tilde{\mathbf{D}}(z) = \frac{z-1}{z-0.95}\begin{bmatrix} 1 & 0 \\ 0 & 1 \end{bmatrix} \tag{6.2}$$

and $\tilde{\mathbf{N}}(z) = \tilde{\mathbf{D}}(z)\mathbf{G}(z)$. Theorem 2 of chapter 2 can be used to show that this factorization is coprime.

Only the **K2** section of the design will be considered below; similar performance may be achieved for command inputs at R. The following constraints were set to limit the control effort, provide robustness for the design, and ensure adequate rejection of low frequency disturbances :

$$|h_{NU}(kT)| \le 50 \qquad 0 \le k \le 50$$

$$|H_{NY}(e^{j2\pi fT})| \le 0.1 \qquad 10 \le f \le 25 \text{ Hz}$$

$$|H_{DY}(e^{j2\pi fT})| \le 0.1 \qquad 0 \le f \le 0.1 \text{ Hz}$$

$$|H_{DY}(e^{j2\pi fT})| \le 2 \qquad 0.5 \le f \le 25 \text{ Hz} \tag{6.3}$$

The cost function to be minimized was based on the time domain response to unit step at the disturbance input D. This in turn gives good decoupling and command tracking performance. For column j the cost is given by

$$J_j = \sum_{k=0}^{50} \sum_{i=1}^{2} a_{i,j}(\mathbf{h}_{DY}[i,j](kT))^2, \quad j \in \{1,2\} \tag{6.4}$$

where

$$a_{i,j} = \begin{cases} 0.1 & i = j \\ 1 & i \ne j \end{cases} \tag{6.5}$$

Elements in the **Q** matrix were approximated as discussed in chapter 2; each had ten coefficients, and the QSTEP parameter was set at four. In figure 6.3 it can be seen that the disturbance rejection constraint has not been met exactly; this is mainly the result of evaluating frequency responses at a finite number of points (50 in this case) on the unit circle. Representing the frequency constraints as linear constraints also gives rise to a small approximation error [30].

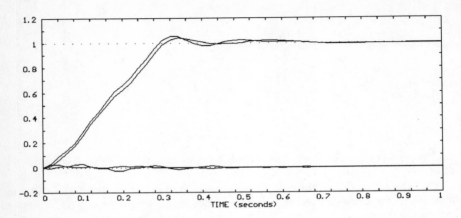

Figure 6.1. Gyroscope h_{NY} step response.

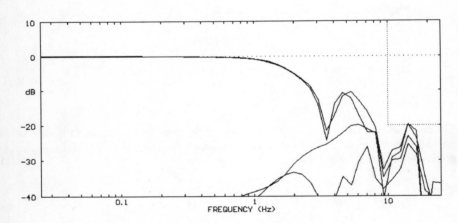

Figure 6.2. Gyroscope H_{NY} frequency response.

Chapter 6. Sample Design Sessions.

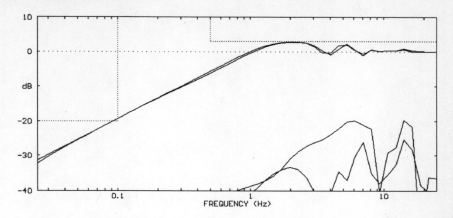

Figure 6.3. Gyroscope H_{DY} frequency response.

The step response of this design (figure 6.1) compares very favourably with that given by Limebeer and Maciejowski [60], and the design displays good robustness properties (figures 6.2 and 6.3).

6.4. Other Examples.

The design package has been applied to numerous other plants. Chapter 7 contains details of the designs for the 2-input 2-output plant used by students during their course on multivariable control systems. A 4-input 4-output model of a minerals processing flotation plant, which included significant dead time, was studied in [47].

6.5. Summary.

Two design examples have been presented in this section. The first example lists the dialogue from a typical design session with the expert system, demonstrating some of the facilities available to assist and guide the designer. The second example illustrates the application of the design method to an unstable plant. In both cases good performance is achieved with a relatively small effort on the part of the designer.

CHAPTER 7
USE OF THE EXPERT SYSTEM

7.1. Introduction.

Educational instruction is another useful application for intelligent design systems [61]. The expert system described here has been used by two groups of students at the University of Cape Town. Although it was not primarily intended for educational purposes, the expert system did prove useful for introducing new concepts to the students. In both instances the students were not familiar with either the design system or design method, and they received minimal explicit tuition on either.

The first group of students, an undergraduate class, used SV-CXS, a very early version of the software, for their control system design project (single variable). The results of this trial were presented in [62], and are summarized below. The second group used the full MV-CXS design system during their postgraduate course on multivariable control system design.

7.2. The Undergraduate Control System Design Project.

The undergraduate control systems design project was the design of a level controller for a sump in a milling process. This process was modeled (and simulated) using a transfer function of the form

$$G(s) = \frac{a}{s} e^{-bs}$$

The students were each given individual values for the parameters a and b, typically 0.2 and 4.0 respectively. Disturbances were added at the input of the plant, and comprised a dc offset, a relatively large disturbance at 0.016Hz plus a number of harmonics, and high frequency noise.

The students performed two designs for this problem. The first was done using the classical methods taught class; generally an open loop Nyquist approach was adopted, resulting in a PI controller. For the second design the expert system was used, with minimal additional tuition.

The performance of a controller for this problem can be judged in terms of its ability to track the setpoint and reject disturbances. Additional considerations are the robustness of the controller, and the demands it places on the actuator. The engineering tradeoff is between a slow, robust, low performance controller, and a sensitive high performance controller.

To compare the response to a step setpoint change, the time to settle to within 5% was taken as the measure of performance. In 89% of the designs the SV-CXS controllers exhibited better responses; the average settling time was 16.2 seconds compared to 63.4 seconds for the classical controller. This was to be expected from the advantage of using a two-parameter control structure over the standard configuration, which also makes the comparison unfair. However it should be remembered that the students would not have been able to design a suitable pre-filter without the expert system.

More important are the disturbance rejection properties. The parameter chosen for the comparison was the frequency below which disturbances at the plant output were rejected by at least 3 dB. The SV-CXS controllers performed better in 70% of the designs in this regard; the average frequency was 0.0146 Hz for the classical design, against 0.0187 Hz for the SV-CXS controller, an improvement of 28%.

Unfortunately almost half of the SV-CXS designs suffered from excessive controller gain or amplification of high frequency disturbances. The failures were due in part to the expert system not adequately warning of the design deficiencies, and partly to the students ignoring the advice that was given. In many cases there was a deliberate decision to tradeoff almost everything for low frequency disturbance rejection.

The results of this experiment were not always as encouraging as had been hoped for, particularly in the sense of the failures mentioned above. Nevertheless there were many positive aspects, and the multivariable expert system "learned" a lot from this experience. Some of the important points highlighted were :

- Many students had difficulty with the two-parameter controller structure. In particular the independence of the setpoint tracking and the disturbance rejection or stability properties of this form was seldom grasped. The expert system could have spent more effort in explaining this, or could have allowed the standard controller structure as an option. Nevertheless the two-parameter structure, together with the optimization facilities, account for the relatively good tracking performance of the SV-CXS controllers.

- While the students were generally comfortable using the Nyquist plot, they were not always able to interpret the closed loop responses. This sometimes resulted in them specifying unreasonable or meaningless performance constraints. While it is possible for the expert system to indicate that a particular specification is questionable, the system has been designed assuming that the designer has a purpose for each specification. It is the designer who needs to make the engineering tradeoffs; the best that the expert system can do is provide advice on what the tradeoffs are, and what the consequences of a certain decision might be.

- Most of the students gained an appreciation of the tradeoffs involved in control system design, often as a result of the poor performance of early designs. A powerful feature of this design method is that it readily allows one to squeeze the specifications on one response and view the effects on the others, identifying explicitly any constraint conflicts.

Chapter 7. Use Of The Expert System.

- Initially the students had almost no conceptual understanding of the optimization of responses. The expert system should have provided a "beginners" guide to optimization, asking the student which features they considered important, and then selecting appropriate optimization weights. The students were impressed by the optimization feature, as they could not do anything similar using the classical techniques they were accustomed to.

7.3. The Postgraduate Control System Design Project.

In this project the students were required to design a controller for the plant

$$\mathbf{G}(s) = \begin{bmatrix} \dfrac{0.1054}{s - 0.1054} & \dfrac{0.3162}{s - 0.1054} \\ \dfrac{0.2108}{s - 0.1054} & \dfrac{0.8924}{s - 0.2231} \end{bmatrix}$$

which satisfied the following specifications on the step response and control signal (assuming unit step inputs):

$$\mathbf{h}_{NY}[i,j](t) \geq 0.95, \quad t \geq 30s, \tag{7.1}$$

$$\mathbf{h}_{NY}[i,j](t) \leq 1.05, \quad t \geq 0s, \tag{7.2}$$

and $\quad |\mathbf{h}_{NU}[i,j](t)| \leq 3, \quad t \geq 0s, \tag{7.3}$

for $i,j \in \{1,2\}$. This plant, a scaled version of that studied by Zhao and Kimura [44], is non-minimum phase, with a right half plane zero near $s = 0.1809$.

The students repeated the design using four different design approaches - initially they used single variable design methods, next the INA/DNA method, the characteristic loci method, and finally the MV-CXS package. Approximately one week was allocated for each design method.

The MV-CXS design was performed on the zero-order hold equivalent of $\mathbf{G}(s)$, sampled at 1 Hz; the transfer function $\mathbf{G}(z)$ is given below. The students received no tuition on using the package other than the instructions given in appendix C.

$$\mathbf{G}(z) = \begin{bmatrix} \dfrac{0.1}{z - 0.9} & \dfrac{0.3}{z - 0.9} \\ \dfrac{0.2}{z - 0.9} & \dfrac{0.8}{z - 0.8} \end{bmatrix}$$

The students used a one parameter controller structure for the classical designs. For the MV-CXS designs, a two parameter structure was assumed; to make the comparison with the other methods fair, only the **K2** section of the design (i.e. inputs N and D) was considered. The students were, however, expected to design for command inputs at input R also.

The designs were evaluated in terms of their ability to reduce interaction and reject disturbances, in addition to satisfying the specifications on the tracking performance and control signal. In the tables below, the column labeled "Satisfied" indicates whether or not the design satisfied the specifications; the "Interaction" column gives the peak off-diagonal response at the plant output to a unit step on the command input. "Bandwidth" is the frequency (Hz) below which all disturbances at any single plant output are rejected by at least 3 dB, and "DYmax" the maximum amplification of any disturbance at the plant output. This last parameter gives an indication of the robustness of the design; the lower the values, the more robust the design should be.

7.3.1. Single Variable Designs.

The students used single variable methods for the first group of designs. For example, some students used a sequential loop closing method, where a controller $k_{11}(s)$ was designed for $g_{11}(s)$, and then a second controller $k_{22}(s)$ was designed for

$$g_{22}(s) - g_{12}(s)k_{11}(s)(1 + g_{11}(s)k_{11}(s))^{-1}g_{21}(s) \tag{7.4}$$

In all cases the single variable designs were done using a classical Nyquist diagram method; in general PI (proportional plus integral) controllers were produced. The two controllers were then combined to give

$$\mathbf{K}(s) = \begin{bmatrix} k_{11}(s) & 0 \\ 0 & k_{22}(s) \end{bmatrix} \tag{7.5}$$

Students (1) and (2) achieved good results by re-assigning the plant inputs (i.e. controlling output 1 using input 2, and output 2 using input 1) based on an analysis of the relative gain array [63]. In subsequent designs (except those using MV-CXS where this assignment is immaterial) almost all students adopted this approach.

The performance of the single variable designs is shown in table 7.1 below. An error in the program computing the performance of the controllers for the single variable designs may have led some students to believe incorrectly that their designs satisfied the requirements; the initial control signal value u(0) was not included when computing the maximum value of the control signal. However the students could have observed this from the graphical plot of the control signal response, and should have been concerned about the corresponding high gains in the disturbance rejection response. These cases have been labeled as (Yes) in the 'Satisfied' column, and will be considered to have satisfied the design requirements.

Chapter 7. Use Of The Expert System.

Table 7.1. Single variable designs.

Student	Satisfied	Interaction	Bandwidth	DYmax
1	Yes	1.26	0.005	5.08
2	Yes	1.02	0.010	3.20
3	No	0.61	0.005	4.04
4	No	0.60	0.010	4.73
5	(Yes)	2.51	0.003	11.90
6	No	3.11	0.004	22.71
7	No	2.23	0.002	9.83
8	(Yes)	2.32	0.003	10.50
9	(Yes)	2.40	0.003	11.08
10		Unstable		
11		Unstable		
12	(Yes)	2.41	0.003	10.58
13	Close	1.27	0.006	4.67

The design by student (2) was judged to be the best in this group. The step response of this design is shown in figure 7.1 below.

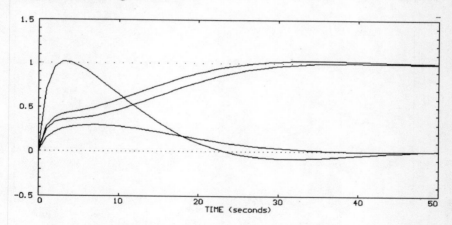

Figure 7.1. Single variable design of student (2).

7.3.2. INA/DNA Designs.

Students were free to use either the Inverse Nyquist Array (INA) or the Direct Nyquist Array (DNA) design method in this section. Some students submitted designs performed using both methods; in these cases the better of the two designs was used in the analysis below. The results of this group of designs are listed in table 7.2.

Table 7.2. INA/DNA designs.

Student	Satisfied	Interaction	Bandwidth	DYmax
1	Close	0.99	0.011	3.28
2	No	1.69	0.004	9.14
3	Close	1.25	0.004	5.62
4	No	0.01	0.008	4.87
5	No	0.55	0	2.97
6	No	0.74	0.013	15.61
7	Yes	1.91	0.005	9.33
8	No	0.50	0.004	7.27
9	Yes	1.53	0.004	7.51
10		Unstable		
11	Close	1.62	0.005	8.66
12		No design		
13	No	1.29	0.010	9.66

The design by student (1) was judged to be the best in this group. The step response of this design is shown in figure 7.2 below.

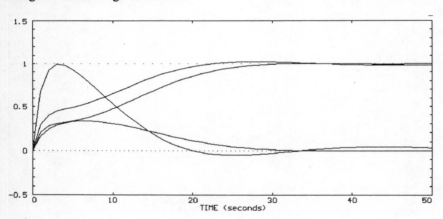

Figure 7.2. INA/DNA design of student (1).

7.3.3. Characteristic Loci Designs.

From table 7.3 it is clear that the Characteristic Loci method produced the least successful results; not one student was able to even nearly satisfy the design requirements using this method. Most students found that while the method gave good indications of stability and interaction, it gave little indication of what should be done to improve the design.

Chapter 7. Use Of The Expert System.

Table 7.3. Characteristic Loci designs.

Student	Satisfied	Interaction	Bandwidth	DYmax
1	No	0.08	0.004	3.38
2	No	0.02	0.004	3.27
3	No	0.70	0.004	8.13
4		Unstable		
5		Unstable		
6		Unstable		
7		No design		
8	No	0.65	0.002	3.72
9	No	0.80	0.005	7.09
10		Unstable		
11	No	2.98	0.002	14.59
12		No design		
13		No design		

The design by student (1) was judged to be the best in this group. The step response of this design is shown in figure 7.3 below.

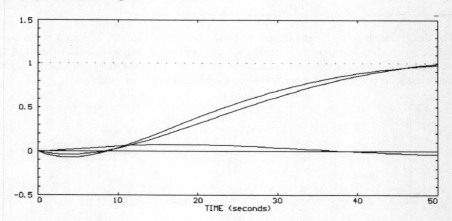

Figure 7.3. Characteristic Loci design of student (1).

7.3.4. MV-CXS Designs.

The students used the MV-CXS design system for the final group of designs. Most students found this easy to use; the NEXT STEP facility appeared to have been particularly helpful. The performance of the controllers is given in table 7.4. The version of MV-CXS used by the students used the first order approximation of Q; had the second order version been available it is most likely that the students would have achieved better results.

All students but one were able to meet the design requirements using MV-CXS. Student (6) did not constrain the off-diagonal elements of the RU response, nor any of those of the NU response. Furthermore, this student required the off-diagonal NY responses to meet the conditions intended for the diagonal elements. However, this student did not achieve even nearly acceptable results using any of the other design methods.

Table 7.4. MV-CXS designs.

Student	Satisfied	Interaction	Bandwidth	DYmax
1	Yes	0.29	0.006	3.89
2	Yes	0.05	0.005	5.14
3	Yes	0.72	0.010	3.32
4	Yes	0.85	0.008	2.17
5	Yes	1.02	0.008	3.27
6	No	1.05	0.000	7.34
7		No design		
8	Yes	0.20	0.006	4.00
9	Yes	0.20	0.005	3.82
10	Yes	0.50	0.008	2.80
11	Yes	No design		
12	Yes	No design		
13	Yes	No design		

The design by student (9) was judged to be the best in this group. The step response of this design is shown in figure 7.4 below; the performance would have been better had the student not placed tight constraints on the undershoot. While the design of student (2) has much lower interaction, it did not achieve asymptotic disturbance rejection.

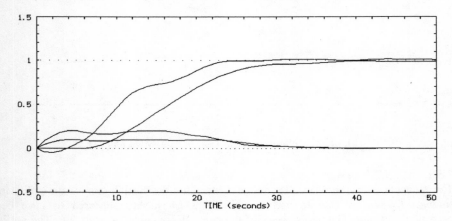

Figure 7.4. MV-CXS design of student (9).

7.3.5. Comparison of Design Methods.
In general it is difficult to compare designs as there are many possible performance measures. However the students were asked to minimize interaction and maximize the disturbance rejection properties of the closed loop system. The designs are therefore

compared on these criteria; in addition the value of the DYmax parameter, which gives an indication of the robustness of the design, is also compared. These criteria are generally complementary; to achieve a lower interaction value it is generally necessary to accept a higher value of DYmax, for example.

The design methods are compared in the tables below; only those designs which satisfied the requirements, or were at least close, are considered. Table 7.5 below gives the method which produced the best results in terms of the interaction, bandwidth and DYmax criteria; table 7.6 gives the average value of each of these criteria. The ratio of designs which satisfied the requirements (at least approximately) to the total number submitted is also given as a percentage in table 7.6. In these tables, S.V. denotes the single variable designs, INA those done using the INA/DNA method, and C.L. the characteristic loci designs.

Table 7.5. Design method giving best performance.

Student	Interaction	Bandwidth	DYmax
1	MV-CXS	INA	INA
2	MV-CXS	S.V.	S.V.
3	MV-CXS	MV-CXS	MV-CXS
4	MV-CXS	MV-CXS	MV-CXS
5	MV-CXS	MV-CXS	MV-CXS
6		Incomplete	
7		Incomplete	
8	MV-CXS	MV-CXS	MV-CXS
9	MV-CXS	MV-CXS	MV-CXS
10	MV-CXS	MV-CXS	MV-CXS
11		Incomplete	
12		Incomplete	
13		Incomplete	

Table 7.6. Average performance of acceptable designs.

Method	Number of acceptable designs	Average interaction	Average bandwidth	Average DYmax
S.V.	7 (54%)	1.89	0.0047	8.14
INA	5 (42%)	1.46	0.0058	6.88
C.L.	0 (0%)	-	-	-
MV-CXS	8 (89%)	0.48	0.0070	3.55

Tables 7.5 and 7.6 above clearly reveal MV-CXS as the most successful of the design methods. Even though the designs involved trading each performance criterion off against the others, the MV-CXS method often produced the best results for all three criteria. However there are some factors which make the comparison with the classical design methods appear mildly unfair. These are considered below.

Firstly the design requirements were posed in the time domain, and the classical design methods operate primarily in the frequency domain. Probably more significant, however, is that MV-CXS allows the designer to satisfy closed loop performance constraints explicitly, while all the other methods can satisfy closed loop requirements only implicitly. Thus had the requirements been set in the frequency domain, it is quite probable that MV-CXS would remain the method of choice. On the other hand the performance constraints were not difficult to satisfy, and two of the three performance criteria used to compare the acceptable designs are frequency domain parameters.

The second possible objection is that the controllers synthesized by MV-CXS are vastly more complex than those of the classical methods. This may be a real concern for many applications; the options for implementing the controllers are discussed in chapter 8. One of those options is to estimate a low order controller; the step response using the controller estimated from the design of student (9) is shown below. Here the elements of the controller transfer function are ratios of second order polynomials. This response is still better than the best of those designed using the classical methods.

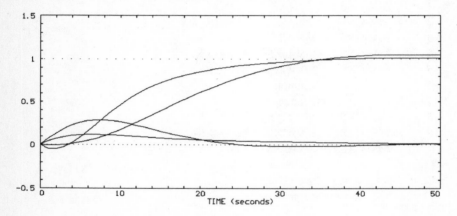

Figure 7.5. MV-CXS design of student (9) with estimated second order controller.

Finally it could be argued that the students had gained a deep understanding of the plant by the time the MV-CXS designs were performed. However, any experience from previous designs is certainly not reflected in the results of the characteristic loci designs. Furthermore, the MV-CXS design method is substantially different from the others, making this possible advantage minor. It is also possible that this experience had a negative effect on the MV-CXS results, in that the students may have stopped improving their designs once they had obtained a design better than their previous attempts. Being last in the sequence also meant that the enthusiasm of the students was at its lowest for this method, as can be seen by the number of students not even attempting the exercise.

On the other hand it should be remembered that the students were given hardly any instruction on the use of MV-CXS, or on the factorization method for control system design. In contrast the classical methods used above were taught in some detail in their classes. Some students, however, had used an early version of the expert system for single variable designs as undergraduates the previous year.

Chapter 7. Use Of The Expert System.

In addition, it is most probable that the MV-CXS design system would have produced even better results relative to the other methods had the dimension of the plant been higher. The relative increase in the level of skill required of the designer as the dimension of the problem increases is far greater for the classical methods than it is for MV-CXS. Using MV-CXS, the designer is simply required to select the most appropriate tradeoffs for the design; the mechanics of the design process remain largely hidden.

In summary, the results of this experiment were extremely encouraging. The goal of the design system, that of assisting users to produce a good control system designs, has been effectively fulfilled. A significant degree of this success is based on the experience gained during the undergraduate design project [62].

7.3.6. Alternative Designs.

In addition to using the design methods above, two students submitted what they called analytic designs. These were based on an algebraic computation of a compensator to give exact decoupling, followed by two single variable designs for the decoupled plant; the first step is not difficult given the simplicity of the plant transfer function. The non-zero interaction values in the tables below arise from imperfections in the simulation program. Of the two designs, that of student (9) was particularly good. The step response of this design is shown in figure 7.6 below. Similar performance is also possible using MV-CXS; figure 7.7 shows a design generated using the second order approximation for Q, with the QSTEP parameter set at 4, and NVARS set at 12.

Table 7.7. Analytical designs.

Student	Satisfied	Interaction	Bandwidth	DYmax
8	Close	0.01	0.005	4.33
9	Yes	0.01	0.006	4.42
MV-CXS	Yes	0.01	0.006	4.42

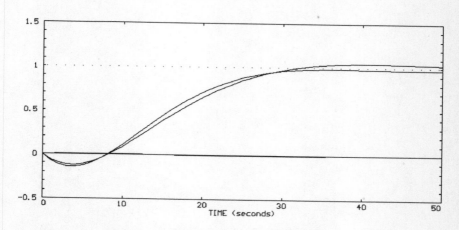

Figure 7.6. Analytical design of student (9).

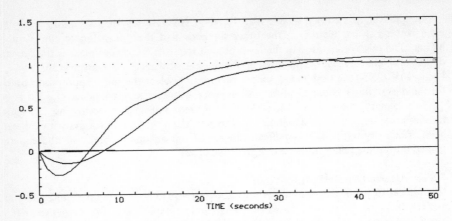

Figure 7.7. MV-CXS design.

7.4. Summary.

The effectiveness of the design system has been tested by two groups of students at the University of Cape Town. An early version, for single-variable design problems, was used by undergraduate students during their control systems design project. This chapter dealt with the use of the MV-CXS design system by postgraduate students during their course on multivariable systems. The results here were particularly encouraging, with the expert system tool enabling them to produce markedly superior results compared to various classical design methods.

CHAPTER 8
IMPLEMENTING THE CONTROLLER

8.1. Introduction.

An unfavourable aspect of the design method is the very high order of the resulting controllers. Boyd *et al.* [30] give three possible uses for controllers produced by this method: they may be implemented in full, reduced order controllers may be derived from them, or they may be used as a standard against which controllers designed by other methods are measured. This chapter examines the first two alternatives.

8.2. Full Controller Implementation.

Boyd *et al.* [30] propose an architecture for implementing the controller directly. However, when the plant transfer function is strictly proper, a much simpler implementation is possible.

Let the initial stabilizing controller K_0 have the stable left-coprime factorization

$$K_0 = Y^{-1}X \tag{8.1}$$

Then equation (2.4) for the controller can be written as

$$K_c = (Y - Q_2\tilde{N})^{-1},$$

$$K_1 = K_c Q_1,$$

and $\quad K_2 = K_c(X + Q_2\tilde{D}). \tag{8.2}$

For clarity it will be assumed that the inputs D, N, and V are all zero. The control signal $u(kT)$ can then be computed as

$$u(kT) = K_c(z)e(kT)$$

$$= [Y - Q_2(z)\tilde{N}(z)]^{-1}e(kT), \tag{8.3}$$

where $\quad e(kT) = Q_1(z)r(kT) - [X(z) + Q_2(z)\tilde{D}(z)]y(kT). \tag{8.4}$

Note that the transfer function matrices Q_1, Q_2, X, and $\tilde{D}$ are stable, and thus the error signal $e(kT)$ at time kT is easily computed from the sequences $\{r(iT), i \leq k\}$ and $\{y(iT), i \leq k\}$. Next it is necessary to find a method for computing K_c.

Partition $Y(z)$ into the real (constant) matrix Y_∞ and the strictly proper $Y'(z)$ such that

$$Y(z) = Y_\infty + Y'(z). \tag{8.5}$$

where

$$Y_\infty = \lim_{z \to \infty} Y(z) \tag{8.6}$$

Next define

$$Y_c(z) = Q_2(z)\tilde{N}(z) - Y'(z) \tag{8.7}$$

and $\quad Y_i = Y_\infty^{-1}. \tag{8.8}$

Assuming that Y_i exists (that is, Y^{-1} exists and is proper or strictly proper), equation 8.3 can be rewritten as

$$(Y_\infty - Y_c(z))u(kT) = e(kT),$$

or $\quad u(kT) = Y_i(e(kT) + Y_c(z)u(kT)). \tag{8.9}$

Note that if the either the plant or controller is stable (factorizations given by equations 2.6 and 2.7 respectively), or the diagonal factorization method discussed in chapter 2 is used to compute the fractions X and Y for K_0 (equation 2.22), then

$$Y_\infty = Y_i = I. \tag{8.10}$$

For strictly proper G, $\tilde{N}$ is also strictly proper, and thus $zY_c(z)$ is proper and can be realized. This gives a direct formula for computing the control signal $u(kT)$ at time kT from the error sequence $\{e(iT), i \leq k\}$, and the sequence $\{u(iT), i < k\}$ representing the control signal history

$$u(kT) = Y_i[e(kT) + (zY_c(z))(z^{-1}u(kT))] \tag{8.11}$$

Thus the controllers may be implemented directly in terms of the stable transfer function matrices Q_1, Q_2, $\tilde{N}$, $\tilde{D}$, and Y_c; figure 8.1 illustrates this architecture. The transfer function matrices Q_1, Y_c, and $(X + Q_2\tilde{D})$, may be approximated by three high order FIR filters as suggested by Boyd *et al.* [30]; note that Y_c, based on $\tilde{N}$ in particular may require a very high order FIR filter for accurate approximation. When the FIR approximation is used, the asymptotic properties of the controller may be preserved by ensuring that the dc gain of the approximation is exactly that of the transfer function form.

Chapter 8. Implementing The Controller.

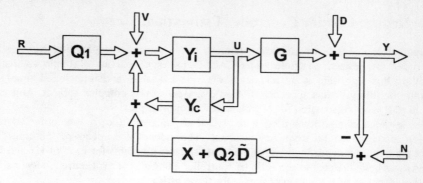

Figure 8.1. Controller structure.

It is of course possible to implement the full controller without using the FIR filter approximation, but implementing $\tilde{N}$, $\tilde{D}$, X and Y' as dynamic systems. Q_1 and Q_2, by their definition, should still be implemented as FIR filters. An efficient structure for this form is shown in figure 8.2. This form simplifies considerably when the plant is stable, as $\tilde{N} = G$, $\tilde{D} = I$, and $X = Y' = 0$.

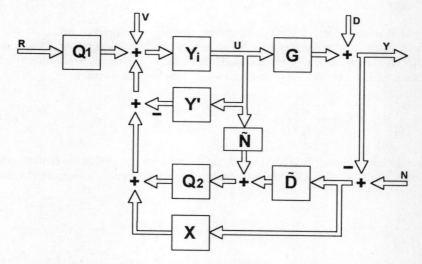

Figure 8.2. Controller structure for full dynamic system implementation.

A consequence of the complexity of the controller is that manual online tuning is virtually impossible; the overall gain of the controller is easily be adjusted, however, by scaling the control signal **u** by a (diagonal) gain matrix. An alternative approach to tuning is "online" re-design of the control system with appropriate modifications to the design specification. Since the design algorithm is fast, this is quite feasible, particularly where the control law can be downloaded directly from the design system to the controller hardware. An expert system could again be profitably employed to assist the user in deciding how to change the design specifications to achieve the desired performance.

8.3. Reduced Order Controller Estimation.

The second approach is to estimate a reduced order controller which approximates the high order synthesized controller. The CACSD package contains a simple estimation algorithm; this algorithm is by no means the best available, and is included simply to illustrate the utility of this approach. For an overview of this complex subject, Anderson and Liu [64] give a discussion of controller reduction principles.

The estimation algorithm used here is a least squares fit of a low order transfer function to the controller frequency response. The individual elements of the controller matrix are treated independently, giving n^2 least squares problems for each of **K1** and **K2**. The analysis below relates to the **K2**[a,b] controller element; the procedure is identical for the other elements of **K2**, and very similar for those of **K1**.

Let the controller element be modeled by the equation

$$k(z) = \frac{n(z)}{d(z)} \qquad (8.12)$$

where $n(z) = n_m z^m + n_{m-1} z^{m-1} + ... + n_0,$ \qquad (8.13)

$d(z) = z^m + d_{m-1} z^{m-1} + ... + d_0,$ \qquad (8.14)

and m is the desired order for the controller element. The original full controller is evaluated at the set of points $\{z_i, i=1...N\}$ on the unit circle in the complex plane, with N > 2m; let K_i denote the controller response **K2**[a,b](z_i). Then the controller estimation can be formulated as a weighted non-linear least squares problem to minimize

$$\sum_{i=1}^{N} w(z_i) \left[K_i - \frac{n(z_i)}{d(z_i)} \right] \qquad (8.15)$$

where $w(z)$ is a frequency dependent weighting function, chosen so that the estimation will be more accurate at critical frequencies. Multiplying each term in the sum by the factor $d(z_i)$ gives

$$\sum_{i=1}^{N} w'(z_i) [K_i d(z_i) - n(z_i)] \qquad (8.16)$$

which is linear in the unknown coefficients. Here the effective weight, in terms of equation 8.15, is

$$w'(z_i) = \frac{w(z_i)}{d(z_i)} \qquad (8.17)$$

Chapter 8. Implementing The Controller.

In this form the coefficients d_i, i=0...m-1, and n_i, i=0...m of the reduced order controller are easily estimated using a standard linear least squares algorithm. Note that the solutions to equations 8.15 and 8.16 are different in general.

The estimation algorithm implemented in the CACSD package is a combination of the linear least squares problems given by equations 8.15 and 8.16, with $w(z) = 1$. First the problem given by 8.16 is solved; the denominator of the controller element is taken from this. A second least squares fit, in the form of equation 8.15 and using the denominator $d(z)$ identified above, is then used to find the numerator polynomial. The corresponding element of the **K1** part of the controller uses the denominator identified for the **K2** part, and solves the equivalent least squares problem given by 8.15 to find the numerator. This approach ensures that the two controller sections have the same poles, which simplifies implementation where the controller has unstable poles (for example at $z = 1$).

8.4. Examples of Controller Implementation.

The first example from chapter 6, the gold mine milling unit with three inputs and three outputs, will be used to illustrate these two approaches to controller implementation. Since the plant is stable, the trivial factorization given by equation 2.6 is used. The implementations are compared using the NY step response, observed over the first 100 seconds.

The response of the full controller is shown in figure 8.3 below.

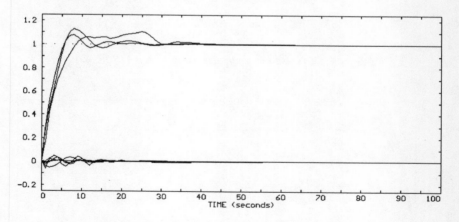

Figure 8.3. Response using the full controller.

8.4.1. FIR Implementation.
The controller structure shown in figure 8.1 is used in this section, with the three dynamic elements of the controller approximated using FIR filters. In this case Y_i is simply the identity matrix, and the FIR filters are the truncated impulse responses of the transfer function matrices

$$\begin{aligned} &\quad Q_1 &&\Rightarrow F_1, \\ &\quad X + Q_2 D = Q_2 &&\Rightarrow F_2, \\ \text{and} &\quad Y_c = Q_2 G &&\Rightarrow F_3. \end{aligned}$$

Of these, the FIR filters for F_1 and F_2 in this example are exactly those of the full order controller; F_3 will differ from Y_c slightly since the impulse response of G is infinite. The effects of this approximation are clearly seen when using 50 tap filters; in particular the integral action has been lost, giving a steady state tracking error.

The responses for the controller implemented using these FIR filters are shown in figures 8.4 and 8.5, using 50 and 75 tap filters respectively.

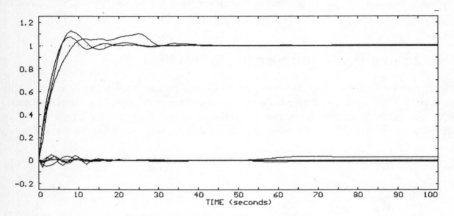

Figure 8.4. Response for controller using 50 tap FIR filters.

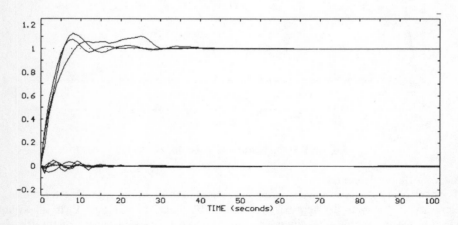

Figure 8.5. Response for controller using 75 tap FIR filters.

Chapter 8. Implementing The Controller.

Attempts to remedy the situation by adjusting the DC gain of the FIR filters to match those of the full order controller are shown in figures 8.6 and 8.7. Again, these plots are for the 50 and 75 tap filters respectively.

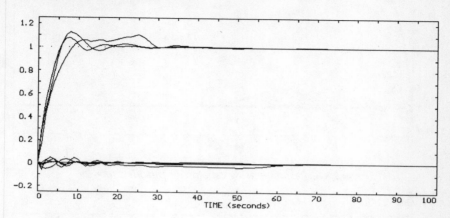

Figure 8.6. Response for controller using 50 tap FIR filters, with DC correction.

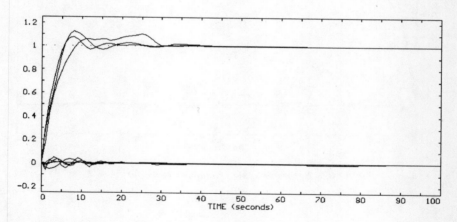

Figure 8.7. Response for controller using 75 tap FIR filters, with DC correction.

8.4.2. Reduced Order Controller Estimation.

Next the estimation method described above was applied to the problem. The estimated controller with first order elements destabilized the system. Figures 8.8, 8.9 and 8.10 show the step response for controllers with elements of order 2, 4 and 6 respectively. It is interesting that, for this example, the step response using fourth order approximation is worse than that using the second order estimate. Better results could be expected using more complex controller estimation techniques.

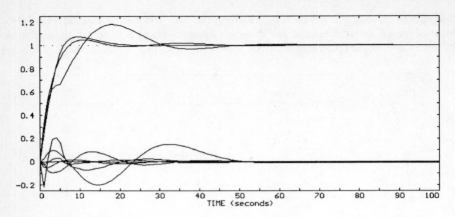

Figure 8.8. Response using estimated controller, m = 2.

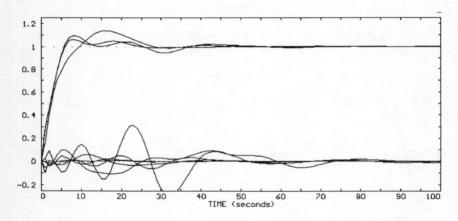

Figure 8.9. Response using estimated controller, m = 4.

Chapter 8. Implementing The Controller.

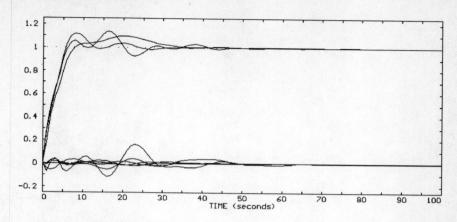

Figure 8.10. Response using estimated controller, m = 6.

8.5. Summary.

Two alternatives for using the complex controllers synthesized by this design method have been discussed in this chapter. The first option is to implement the controllers in full, or at least approximately using high order finite impulse response filters. The second alternative is the estimation of reduced order controllers. A simple identification algorithm which estimates a low order controller from the frequency response of the full controller has been used to demonstrate the feasibility of this approach.

CHAPTER 9
CONCLUSIONS

A powerful CACSD package, based on the design method of Boyd *et al.* [30], has been constructed. This method translates the control system design problem into, and solves it as, a linearly constrained quadratic programming problem. The efficiency of the design method, in terms of both memory requirements and execution speed, has been improved substantially. To achieve this, a diagonal factorization technique has been developed; when applied to the left factorization of the plant transfer function matrix, it allows the multivariable design problem of size αN^2 to be reduced to N sub-problems each of size αN, which may then be solved independently. Although this diagonal factorization is not always coprime, it is suitable for a wide range of plant transfer function matrices. A theorem to check that the factorization is coprime was developed, and is easy to apply. Formulae for (non-diagonal) coprime factorizations, where the nominal stabilizing controller is stable, have also been presented. A novel parameterization for the design transfer function **Q** has been introduced; by appropriate choice of the QSTEP parameter the designer benefits from the efficiency of a low order approximation and while often enjoying almost the same precision as for a high order approximation. Finally, a very efficient representation for the linear constraints generated by the design method has been developed.

The extent of these improvements in efficiency is such that the solution of substantial multivariable control system design problems is practical using only a microcomputer. Similarly, these techniques would enable large scale problems to be tackled on a more powerful workstation.

The complex controllers synthesized by this design method may be implemented in full, or using high order finite impulse response filters. Controller reduction provides a further alternative. A simple identification algorithm to estimate a low order controller from the frequency response of the full controller has been included in the CACSD package, and used to demonstrate the feasibility of this approach.

An expert system interface to this CACSD package has been implemented to produce an intelligent, interactive design tool. The expert system guides and assists both novice and experienced designers in using the CACSD package. Based upon a database of common design features, the NEXT STEP and SUGGEST commands are able to guide and assist the user in formulating and refining the design specification. Similarly the COMPLETE command helps the user to check that the design is complete. Through analysis of the active set and Lagrange multipliers from the quadratic programming problem, the expert system is able to assist the user in identifying and dealing with conflicting performance constraints. The expert system has also been used to effectively extend the scope of the design method, as well as to integrate information from analyses of the design.

The expert system has been implemented on a personal computer, co-resident in memory with the CACSD package. This combination has produced a comprehensive control system design tool, which is nevertheless easy to use. Using a more powerful computer would allow the knowledge base to be extended further, covering even more design situations.

The effectiveness of the design system has been tested by students at the University of Cape Town. An early version, for single-variable design problems, was used by undergraduate students during their control systems design project; the complete MV-CXS design tool was used by postgraduate students during their course on multivariable systems. In both cases, the expert system was used to introduce the students to a new approach to control system design, with minimal additional tuition. The results for the multivariable designs were particularly encouraging, with the expert system tool enabling the students to produce markedly superior results when compared to various classical design methods.

Despite the power of the design system there remains a chasm between artificial and real intelligence. Since the designer has real intelligence, the aim of the expert system has been to assist and complement, rather than replace, the designer. This produces a team solution, which capitalizes on the inherent strengths of both the designer and computer.

APPENDIX A
MV-CXS SPECIFICATIONS

A.1. Requirements for the Computer.

The MV-CXS design system runs on an IBM-PC or compatible computer, under the MS-DOS operating system (version 2.0 or later). At least 640k bytes of RAM, and a 80x87 floating point coprocessor, are required. Hercules, CGA, EGA or VGA graphics facilities are necessary to view the graphics plots.

A.2. Requirements for the Plant.

The plant must be described by a linear, time-invariant, z domain transfer function matrix. Each element in this matrix must be a rational function with real coefficients, and must be strictly proper; the order of the numerator and denominator polynomials must not exceed 50, and they must be coprime. The plant may have up to 5 inputs and 5 outputs.

If the plant is unstable, a stabilizing controller is required. This controller should be stable if possible. MV-CXS produces a two parameter controller, with the configuration shown in figure A.1.

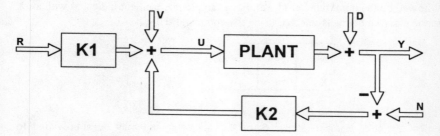

Figure A.1. Two parameter controller configuration.

A.3. The MV-CXS Command Language Specification.

After the initial loading of the plant transfer function and previous design (if any), the MV-CXS design session is mainly command driven. The commands listed below may be used during the design phase.

A.3.1 Commands to Select a Response.

Many commands relate to a particular response. The response currently selected is shown by the command prompt. For example the prompt

```
NY[2,3](time) >
```

indicates that element [2,3] of the response to a unit step at the N input, as observed at output Y, is selected. The commands listed below are used to change the selected response.

RY, RU, NY, NU, or DY
These are used to select a closed loop response. Commands such as EDIT and PLOT, for example, refer to the currently selected response. Note that the first letter refers to the input node for the response, for example R, and the second to the output node, for example Y. The two parameter controller configuration is shown in figure A.1.

VY, or VU
Similar to those above, except that only the frequency responses may be viewed. It is not possible to place constraints on these responses, although constraints on the singular values are allowed.

G, GK, K1, or K2
Also similar to those above, but for the open loop responses. Note that GK refers to GK2. Step responses are not available for GK, K1 or K2.

i j
To select a specific element of a response, enter the output number (1...n), followed by the input number (1...n). For example, the element indicated with an X in the diagram below is selected using the command '3 1'.

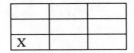

TIME
Select the time domain (step) response. This command may be abbreviated to simply T.

FREQUENCY
Select the frequency response. This command may be abbreviated to FREQ or simply F.

Appendix A. MV-CXS Specifications.   111

A.3.2 Graphics Commands.
The commands listed below are used to display many types of responses graphically. The vertical scale of the graphs may be changed by typing S once the graph has been displayed; for the Nyquist type plots, this automatically adjusts the horizontal scale also.

PLOT
Plot the currently selected response element. For example,
      NY[2,3](time) >**PLOT**
will display the response at output Y2 following a unit step at input N3.

GROUP PLOT
Plot all elements of the currently selected response. The command may be abbreviated to GP.

SVD
Plot the maximum and minimum singular values of the currently selected frequency response.

NYQUIST
Display the Nyquist plot of the currently selected frequency response. The command may be abbreviated to NYQ.

NYQUIST ARRAY
Display the array of Nyquist plots for the currently selected frequency response. The command may be abbreviated to NA.

DNA
Plot the Direct Nyquist Array of the **GK2** frequency response, with Gershgorin circles.

INA
Plot the Inverse Nyquist Array of the **GK2** frequency response, with Gershgorin circles.

A.3.3 Commands to Enter the Specification.
The commands below usually refer to the currently selected response element, for example the DY[1,1] frequency response. Some of the commands may also relate to groups of elements; valid groups are :

ALL	all elements of the current response.
DIAG	the elements on the diagonal of the current response.
OFFDIAG	the elements not on the diagonal of the current response.
COLUMN	the elements in the current column of the current response.

EDIT [group]
Edit the performance constraints on the current response element, or group of elements. This command brings up a spreadsheet style editor. Use the cursor keys (with or without the CTRL key) to select the fields, and CTRL-Y to delete a field. When no value is given in the "value" field, the constraint is assumed void. Zero is assumed when the "from" field is blank, and the range maximum is assumed when the "to" field is blank.

EDIT SVD
As above, but for constraints on the singular values of the frequency response.

SET OPT weight [group]
Set the optimization weight for the current response element, or group of elements. In the time domain the effective cost function is

$$J = \sum_{k=0}^{N} \left[\text{weight}(kT) \times \text{error}^2(kT) \right]$$

where N is the number of samples. For the frequency responses the cost is approximately

$$J = \int_0^{\pi} (\text{weight}(\omega) \times \text{error}^2(e^{j\omega})) \, d\omega$$

In both cases, error is computed as

$(I - h_c)$ or $(I - H_c)$, $\quad c \in \{RY, NY\}$

or

h_c or Hc $\quad$ elsewhere.

RESET OPT [group]
Cancel any optimization on the current response element, or group of elements.

OPT TYPE n
Sets the type of optimization used for subsequent optimization specifications, by modifying the weight specified in the SET OPT command. If a frequency response is currently selected, then this command applies to the optimization of frequency responses; similarly to set the type for the step responses, first select any time domain response. For time domain responses the possible types (effective weights) are :
$\quad$ n=1 : weight
$\quad$ n=2 : weight $\times$ (kT)
$\quad$ n=3 : weight $\times$ (kT)^2
where k is the sample number, and T the sample time. For the frequency domain responses they are :
$\quad$ n=1 : weight
$\quad$ n=2 : weight $\times \omega$

n=3 : weight / ω

where ω is the argument of the frequency when mapped to the unit circle (0 to π radians).

SHOW OPT
Display the elements of the current response which have optimization specified.

SHOW OPT ALL
Display the responses which have optimization specified.

SET ASYM [group]
Set the asymptotic properties for the current response element, or group of elements. For the RY response, the dc response will be forced to the identity matrix, and for DY to the zero matrix. Asymptotic properties may not be set on other responses.

RESET ASYM [group]
Cancel any asymptotic specification on the current response element, or group of elements.

SET ASYM TRACKING
Equivalent to SET ASYM ALL for the RY response.

RESET ASYM TRACKING
Equivalent to RESET ASYM ALL for the RY response.

SET ASYM REJECTION
Equivalent to SET ASYM ALL for the DY response.

RESET ASYM REJECTION
Equivalent to RESET ASYM ALL for the DY response.

SHOW ASYM
Display the elements of the current response which have asymptotic specifications.

SHOW ASYM ALL
Display those responses which have asymptotic specifications.

SAMPLES [n]
Set the value of the SAMPLES parameter to n. If n is omitted, display the current value of SAMPLES. Step responses are evaluated and displayed over the time period 0 to SAMPLES*T, where T is the sample time. The maximum value for SAMPLES is 100.

NFREQ [n]
Set the value of the NFREQ design parameter to n. If n is omitted, display the current value of NFREQ. NFREQ is the number of points used to evaluate frequency responses. The maximum value for NFREQ is 50.

NVARS [n]
Set the value of the NVARS design parameter to n. If n is omitted, display the current value of NVARS.

QSTEP [n]
Set the value of the QSTEP design parameter to n. If n is omitted, display the current value of QSTEP.

A.3.4 Commands to Assist the Designer.
The expert system is programmed with a large amount of information to assist the designer. Besides the commands listed below, the user may also type /EXPLAIN to obtain further help on the specific question being asked.

HELP
This command invokes a menu driven help system, covering the use of the design package as well as guidance for formulating the specification.

HELP topic
This gives help on specific topics. The topics available are SELECT, EDIT, OPT, ASYM, SOLVE, PLOT, SVD, NYQUIST, INA, DNA, SAVE, ESTIMATE, SAMPLES, NFREQ, QSTEP, and NVARS.

NEXT STEP
Execute a step by step design mode. This command, which may be abbreviated to NS, is used to initiate and continue the process.

SUGGEST
This command may be used to provide assistance with formulating the specification, and dealing with conflicting constraints.

SUGGEST EDIT
This command provides information on how the currently selected response may be constrained to improve the control system's performance.

SUGGEST OPT
This command provides assistance with the optimization facilities, and on how they may be used to improve the control system's performance.

COMPLETE
This command is used to help check that the design is complete.

Appendix A. MV-CXS Specifications.

A.3.5 Miscellaneous Commands.

SOLVE
Find a controller, if possible, which meets the current specifications.

ESTIMATE n
Estimate a controller of order n. Subsequent to this command, the various plot commands will show the response using the reduced order controller. If $n < 0$, then revert to the full controller.

SHOW K1
SHOW K2
Display the transfer function of the estimated controller (the **K1** or **K2** section).

SAVE
Save the current design.

REPORT
Print a report on the design.

EXIT
Terminate the design session.

APPENDIX B
THE CACSD PACKAGE INTERFACE

The interface between the expert system and the CACSD package is specified below. The CACSD package is memory resident, and is organized as a library of functions which the expert system can call upon.

The expert system uses the INTR statement to communicate with the CACSD package. This statement has the syntax

```
INTR intr,P1,P2
```

where INTR is the interrupt number to used (96 in this application), and P1 and P2 are two integer parameters. In general, P1 specifies the function required, and P2 the sub-function. Additional data is transferred through the array of expert system variables T0, T1, ... T99.

Some functions refer to a specific response, for example RY[2,3](time); for these the response must first be set up using function 35. Functions with $P1 \geq 1000$ use the currently selected response, and those with $P1 \geq 2000$ use the current element of that response. If $P1 \geq 3000$, then the appropriate element data structure is first created if necessary.

Table B.1 below lists the CACSD functions and their parameters. The parameters marked '⇨' are sent from the expert system to the CACSD package, and those marked '⇦' are returned by the package to the expert system. Some of the functions refer to a file name; the file name is generally made up of the project name (for example MILL), followed by the file name extension (for example .GZ). The code numbers used for the file name extensions are listed in table B.2.

Table B.1. The CACSD functions.

P1	P2	Description
1	x	Edit matrix x (x = 0 : plant transfer function)
		(x = 1 : nominal controller transfer function)
		⇨ T0 = access code
		0 = read only
		1 = edit transfer functions
		2 = (1) and change dimensions
		⇦ T0 = dimension

Table B.1. The CACSD functions (continued).

P1	P2	Description
2	x	Save matrix x (x as above)
3	n	Set optimization type n for time domain responses.

4		Define the CXS symbol values.
		⇨ T0 = unknown
		  T1 = RY
		  T2 = RU
		  T3 = DY
		  T4 = NY
		  T5 = NU
		  T6 = time
		  T7 = frequency
		  T8 = max
		  T9 = min
		  T10 = active
		  T11 = satisfied
		  T12 = unsatisfied
		  T13 = GK
		  T14 = G
		  T15 = K1
		  T16 = K2
		  T17 = VY
		  T18 = VU

3005		Edit constraints ⇦ T5 = changed (0/1) T6 = some constraints (0/1)
1006		Group plot
2006		Plot response
7		Display amount of memory still available

8	Qn	Solve (for **Q1** or **Q2**)
		⇨ T0 = column to solve for (1 - max)
		⇦ T0 = asymptotic constraint conflict (0/1)
		  T1 = constraint conflict (0/1)
		  T2 = solved (0/1)
		  T3 = interrupted (0/1)
		  T4 = asym ratio

Appendix B. The CACSD Package Interface.

Table B.1. The CACSD functions (continued).

P1	P2	Description
2009		Get optimization values ⇐ T5 = weight
3009		Set optimization values ⇒ T5 = weight
10	0 1	Direct Nyquist Array plot (of **GK2**). Inverse Nyquist Array plot (of **GK2**).
2011		Get asymptotic value ⇐ T5 = asymptotic value
3011		Set asymptotic value ⇒ T5 = asymptotic value
12		Save design in file "project.DSN"
13		Load design in file "project.DSN"
15		Set n_vars parameter ⇒ T0 = n_vars
1016		Display Nyquist array.
2016		Display single Nyquist plot.
17	i	Compute minimum distance from zeros of $\tilde{D}[i,i]$ to point $z = 1$. ⇐ T0 = distance
18		Estimate a controller ⇒ T0 = order ⇐ T0 = sufficient memory (0/1)

19	0
	1
		Evaluate H0/1/2 using initial **K0**
		Evaluate H0/1/2 using estimated **K**
		⇐ T0 = status
		    -2 : insufficient memory
		    -1 : no stabilizing controller given
		     0 : plant stable
		     1 : right coprime factorization found
		     2 : **K0** stable
		     3 : no right coprime factorization found and **K0** not stable.

Table B.1. The CACSD functions (continued).

P1	P2	Description
20		Evaluate maximum modulus of plant poles. ⇐ T0 = maximum pole modulus
1021		Plot singular values (SVD) of response.
2022		Get constraint status for current response ⇐ T5 = status (satisfied / active / unsatisfied / unknown)
23		Release memory allocated for controller estimation. Use full order controller.

2024		Examine response over range (time or frequency)
		⇨ T5 = range low end
		    T6 = range high end
		⇐ T7 = minimum value over range
		    T8 = maximum value over range

1025	0 1	Examine SVD minimum conditions set Examine SVD maximum conditions set

2025	0
	1
		Examine minimum conditions set
		Examine maximum conditions set
		⇨ T5 = range low
		    T6 = range high
		⇐ T7 = max over range
		    T8 = min over range
		    T9 = complete (0/1)

26		Get project name ⇐ T0 = project name (far ptr)
27	ext	Get file name (see table B2 for .ext codes) ⇐ T0 = filename = "project.ext" (far ptr)
28	ext	Test if file exists ⇐ T0 = exists("project.ext") (0/1)
29		Set QSTEP parameter ⇨ T0 = QSTEP
30		Set T_SAMPLE parameter ⇨ T0 = T_SAMPLE

Appendix B. The CACSD Package Interface. 121

Table B.1. The CACSD functions (continued).

P1	P2	Description
2031		Copy conditions on the current element ⇨ T2 = destination input T3 = destination output
32		Release memory allocated for **H0/H1/H2** tables.
33		Check plant transfer function element ⇨ T0 = output number T1 = input number ⇦ T2 = strictly proper (0/1) T3 = coprime (0/1)
34		Perform left MFD (matrix fraction decomposition) of plant ⇦ T0 = $\tilde{D}$ equals **I**, i.e. plant is stable (0/1) T1 = left MFD is coprime (0/1)
35		Select response as current ⇨ T0 = domain (Time/Frequency) T1 = response name (e.g. RY) T2 = input number T3 = output number
36	0	Check that controller satisfies performance constraints on current response.
	1	Check that controller satisfies asymptotic constraints on current response. ⇦ T0 = satisfied (0/1)
37		Check for (and then reset) memory allocation failures ⇦ T0 = local allocation failure T1 = far allocation failure
1038	x	Edit SVD constraints (if x≠0, edit the minimum constraints also) ⇦ T5 = changed (0/1) T6 = some constraints (0/1)
1039		Test SVD constraints. ⇦ T5 = satisfied (0/1)
1040		Check if any SVD constraints ⇦ T5 = some (0/1)

Table B.1. The CACSD functions (continued).

P1	P2	Description
41		Set variable MAX_POLE_MOD ⇨ T0 = MAX_POLE_MOD
42	0	Save the **Q** matrix
	1	Load the **Q** matrix
43		Print report
44		Set parameter NFREQ ⇨ T0 = NFREQ
45		Set parameter samples ⇨ T0 = samples
46		Initialize the **Q1** and **Q2** matrices.
47		Copy **K2** to **K0**.
48	n	Set optimization type n for frequency domain responses.

Table B.2. Filename extension codes.

Code	Extension
0	.GZ
1	.K0Z
2	.PRM
3	.DSN
4	.DBF
5	.Q
6	.SPC
7	.SOL
8	.K1Z
9	.K2Z

APPENDIX C

THE MV-CXS STUDENT DESIGN PROJECT INSTRUCTIONS

Using the MV-CXS Design Package.
An expert system design package, running under the MV-CXS expert system, has been developed to assist the user with multivariable control system design problems. It is based upon a new design method, where specifications on the closed loop system are satisfied explicitly.

Starting the Design System.
Make your own copy of the MV-CXS disk, as your design parameters and specifications will be saved on it. Insert this disk into the A drive, and enter the following commands :

```
C:>A:
A:>HARDCOPY           (If you want to print graphics)
A:>GO
```

The system will now start loading. Do not remove the disk until the design session is over.
  To print graphics screens, press the SHIFT and PrtSc keys simultaneously, and then press the 1 key. Remember that you must have run the HARDCOPY program first.

Using the Expert System Effectively.
The expert system is programmed with many features to assist you with your design. For example, if you do not understand a question, type **/EXPLAIN** (or **/E** for short) to get a more detailed explanation of it.
  Most of the design session is driven by commands. Details of the commands available can be seen by typing the **HELP** command. Information on a specific command, such as **EDIT**, can also be obtained by typing **HELP EDIT**.

The structure used for the control system is shown below. Nodes R, V, D and N are inputs, and U and Y are outputs. Note that N is equivalent to the command input for the standard controller configuration (i.e. **K2** only).

The expert system refers to responses using a two letter name; for example DU indicates the response from input D to output U. In addition a particular element of the response matrix, for example [2,2], can also be selected. A typical prompt, showing the currently selected response and domain, is

    RY[2,1](time) >

This indicates that commands such as **PLOT** will operate on the RY step response element [2,1] (i.e. the response to a unit step at input R1, as observed at output Y2). The current response may be changed as desired (type **HELP SELECT** for details). Note that it is also possible to select certain open loop frequency responses; these are G, GK, K1 and K2.

The overall design sequence is to enter the design specification using the **EDIT**, **ASYM**, and **OPT** commands. Then use the **SOLVE** command to find a controller (if possible) which meets these specifications. Check the responses with the **PLOT** command. If necessary, revise the specifications using the **EDIT** command again. There are two other commands to help you: **NEXT STEP**, or **NS** for short, guides you step by step in developing the specification, and **SUGGEST** can provide further suggestions on what to do.

Objectives for the Design.
The design must satisfy the following specifications :

- The step response of the diagonal elements of NY and RY (the command inputs) must rise to at least 95% within 30 seconds. The overshoot must not be more than 5%.

- The absolute value of any of the control signals U must not exceed 3 units following a unit step input at R or N.

In addition to these, the interaction seen after step inputs at R or N should be minimized, as should the maximum amplification of disturbances at input D.

Report.
Once the design is complete, the **REPORT** command must be used to document the design specifications and performance. Be sure that the printer is ready!

REFERENCES

1. Åström KJ et al., "Expert Control", *Automatica*, 1986, vol. 22, no. 3, pp. 277-286.

2. Antsaklis PJ, "Neural Networks in Control Systems", *IEEE Control Systems Magazine*, 1990, vol. 10, no. 3, pp. 3-5.

3. Haest M et al., "ESPION: an Expert System for System Identification", *Automatica*, 1990, vol. 26, no. 1, pp. 85-95.

4. Larsson JE, Persson P, "Knowledge Representation by Scripts in an Expert Interface", *Proceedings of the American Control Conference*, Seattle, USA, 1986, pp. 1159-1162.

5. Graham N, *Artificial Intelligence*, Tab Books, Blue Ridge Summit, PA, 1979.

6. Rich E, *Artificial Intelligence*, McGraw-Hill, New York, 1983.

7. Denham MJ, "Design Issues for CACSD Systems", *Proceedings of the IEEE*, 1984, vol. 72, no. 12, pp. 1714-1723.

8. Dreyfus HL, Dreyfus SE, *Mind over Machine*, The Free Press, New York, 1986.

9. Dreyfus HL, Dreyfus SE, "Why Expert Systems Do Not Exhibit Expertise", *IEEE Expert*, 1986, vol. 1, no. 2, pp. 80-83.

10. Pang GKH, MacFarlane AGJ, *An Expert Systems Approach to Computer-Aided Design of Multivariable Systems*, Springer-Verlag, Berlin, 1987.

11. Jamshidi M, Herget CJ, eds., *Computer-Aided Control Systems Engineering*, North-Holland, Amsterdam, 1985.

12 Taylor JH, Frederick DK, "An Expert System Architecture for Computer-Aided Control Engineering", *Proceedings of the IEEE*, 1984, vol. 72, no. 12, pp. 1795-1805.

13 Åström KJ, "Computer Aided Tools for Control System Design". In Jamshidi M, Herget CJ (eds), *Computer-Aided Control Systems Engineering*, North-Holland, Amsterdam, 1985, pp. 3-40.

14 Pang GKH *et al.*, "Development of a New Generation of Interactive CACSD Environments", *IEEE Control Systems Magazine*, 1990, vol. 10, no. 5, pp. 40-44.

15 Buchanan BG, Shortliffe EH, *Rule Based Expert Systems*, Addison-Wesley, 1984.

16 Hayes-Roth F *et al.*, *Building Expert Systems*, Addison-Wesley, Reading MA, 1983.

17 Michie D (ed), *Introductory Readings in Expert Systems*, Gordon and Breach Science Publishers, New York, 1982.

18 Rychener MD, *Expert Systems for Engineering Design*, Academic Press, New York, 1988.

19 James JR *et al.*, "A retrospective view of CACE-III: considerations in co-ordinating symbolic and numeric computation in a rule based expert system", *Proceedings of the Second Conference on A.I. Applications*, Miami Beach, Florida, 1986.

20 James JR *et al.*, "Use of expert-systems programming techniques for the design of lead-lag compensators", *IEE Proceedings Part D*, 1987, vol. 134, no. 3, pp. 137-144.

21 Edmunds JM, "Cambridge linear analysis and design program", *Proceedings of the 1st IFAC Symposium on Computer-Aided Control System Design*, Zurich, 1979, pp. 253-258.

22 Trankle TL *et al.*, "Expert System Architecture for Control System Design", *Proceedings of the American Control Conference*, Seattle, USA, 1986.

23 Little JN *et al.*, "CTRL-C and Matrix Environments for the Computer-Aided Design of Control Systems". In Jamshidi M, Herget CJ (eds), *Computer-Aided Control Systems Engineering*, North-Holland, Amsterdam, 1985, pp. 111-124.

References.

24 Nolan PJ, "An Intelligent Assistant for Control System Design", *Proceedings of the 1st International Conference on the Application of Artificial Intelligence in Engineering Problems*, University of Southampton, U.K., 1986.

25 Birdwell JD *et al.*, "Expert Systems Techniques in a Computer-based Control System Analysis and Design Environment", *Proc. 3rd IFAC/IFIP Symposium*, Lyngby, Denmark, 1985.

26 Birdwell JD, "An Expert System can aid in the Evolution of a Design Methodology", *Proceedings of the American Control Conference*, 1987.

27 Pang GKH, "An Expert System for CAD of Multivariable Control Systems using a Systematic Design Approach", *Proceedings of the American Control Conference*, 1987.

28 Boyle JM *et al.*, "The development and implementation of MAID: a knowledge based support system for use in control system design", *Transactions of the Institute of Measurement and Control*, 1989, vol. 11, no. 1, pp. 25-39.

29 MacFarlane AGJ *et al.*, "Future design Environments for Control Engineering", *Automatica*, 1989, vol. 25, pp. 165-176.

30 Boyd SP *et al.*, "A New CAD Method and Associated Architectures for Linear Controllers," *IEEE Transactions on Automatic Control*, 1988, vol. AC-33, no. 3, pp. 268-283.

31 Boyd SP *et al.*, "Linear Controller Design : Limits of Performance via Convex Optimization", *Proceedings of the IEEE*, 1990, vol. 78, no. 3, pp. 529-574.

32 Fegley KA, "Designing Sampled-Data Control Systems by Linear Programming", *IEEE Transactions on Applications and Industry*, 1964, vol. 83, pp. 198-200.

33 Gustafson CL, Desoer CA, "Controller design for linear multivariable feedback systems with stable plants, using optimization with inequality constraints", *International Journal of Control*, 1983, vol. 37, no. 5, pp. 881-907.

34 Gustafson CL, Desoer CA, "A CAD methodology for linear multivariable feedback systems based on algebraic theory", *International Journal of Control*, 1985, vol. 41, no. 3, pp. 653-675.

35 Becker RG *et al.*, "Computer-aided design of control systems via optimization", *Proceedings of the IEE*, 1979, vol. 126, no. 6, pp. 573-578.

36 Polak E *et al.*, "Control System Design Via Semi-Infinite Optimization: A Review", *Proceedings of the IEEE*, 1984, vol. 72, no. 12, pp. 1777-1794.

37   Zakian V, Al-Naib U, "Design of dynamical and control systems by the method of inequalities", *Proceedings of the IEE*, 1973, vol. 120, no. 11.

38   Bhattacharyya SP *et al.*, "Stabilizability conditions using linear programming", *IEEE Transactions on Automatic Control*, 1988, vol. 33, no. 5, pp. 460-463.

39   Moore KL, Bhattacharyya SP, "A technique for choosing zero locations for minimal overshoot", *IEEE Transactions on Automatic Control*, 1990, vol. 35, no. 5, pp. 577-580.

40   Vidyasagar M, *Control System Synthesis: A Factorization Approach*, M.I.T. Press, Cambridge, MA, 1985.

41   Rosenbrock HH, *Computer-Aided Control System Design*, Academic Press, New York, 1974.

42   Youla DC *et al.*, "Modern Wiener-Hopf design of optimal controllers, part II: The multivariable case", *IEEE Transactions on Automatic Control*, 1976, vol. 21, no. 3, pp. 319-338.

43   Nett CN *et al.*, "A Connection Between State-Space and Doubly Coprime Fractional Representations," *IEEE Transactions on Automatic Control*, 1984, vol. AC-29, no. 9, pp. 831-832.

44   Zhao Y, Kimura H, "Two-degree-of-freedom dead-beat control system with robustness: multivariable case", *International Journal of Control*, 1989, vol. 49, no. 2, pp. 667-679.

45   Francis BA, *A Course in H^∞ Control Theory*, Springer-Verlag, Berlin, 1987.

46   Kailath T, *Linear Systems,* Prentice-Hall, Englewood Cliffs, NJ, 1980.

47   Tebbutt CD, "A Microprocessor Implementation of Multivariable Factorization Theory", *IEEE Transactions on Automatic Control*, 1992, vol. 37, no. 10, pp. 1631-1634.

48   Tebbutt CD, "An Efficient Representation for Linear Constraints", *IEEE Transactions on Automatic Control*, 1990, vol. 35, no. 8, pp. 949-951.

49   Scales LE, *Introduction to Non-Linear Optimization*, MacMillan Publishers, London, 1985.

50   Golub GH, van Loan CF, *Matrix Computations*, North Oxford Academic, Oxford, 1983.

References.

51	Businger PA, Golub GH, "Singular Value Decomposition of a Complex Matrix", *Communications of the ACM*, 1964, vol. 12, no. 10, pp. 564-565.
52	Tebbutt CD, "Expert Systems Approach to Controller Design", *IEE Proceedings Part D*, 1990, vol. 137, no. 6, pp. 367-373.
53	Tebbutt CD, "An Expert System for Controller Design", *Transactions of the SAIEE*, 1990, vol. 81, no. 3, pp. 15-19.
54	Genesereth MR, "The role of plans in intelligent teaching systems". In Sleeman D and Brown JS (eds), *Intelligent Tutoring Systems*, Academic Press, London, 1982.
55	Jackson P, Lefrere P, "On the application of rule-based techniques to the design of advice-giving systems". In Coombs MJ (ed), *Developments in expert systems*, Academic Press, London, 1984.
56	Winston PH, Horn BKP, *Lisp (2nd edition)*, Addison-Wesley, Reading, MA, 1984.
57	Doyle JC, Stein G, "Multivariable Feedback Design: Concepts for a Classical/Modern Synthesis", *IEEE Transactions on Automatic Control*, 1981, vol. 26, no. 1, pp. 4-16.
58	Tebbutt CD, *The CXS Expert System Shell*, 1990.
59	Hulbert DG, Braae M, "Multivariable Control of a Milling Circuit at East Driefontein Gold Mine", *National Institute for Metallurgy*, 1981, Report 2113.
60	Limebeer DJN, Maciejowski JM, "Two tutorial examples of multivariable control system design", *Transactions of the Institute of Measurement and Control*, 1985, Vol. 7, no. 2, pp. 97-107.
61	Birdwell JD *et al.*, "Teaching with the CASCADE Computer-Aided Control System Design Environment", *Proceedings of the 19th Southeastern Symposium on System Theory*, Clemson SC, 1987.
62	Tebbutt CD, "Use of an Expert System for Controller Design", *presented at the RUGSA Symposium on the Applications of Knowledge-based Systems in Engineering and Control*, Johannesburg, South Africa, 1989. Reprinted in *Computech*, 1990, vol. 6, no. 7, pp. 30-33.
63	Maciejowski JM, *Multivariable Feedback Design*, Addison-Wesley, Wokingham, England, 1989.

64   Anderson BDO, Liu Y, "Controller Reduction : Concepts and Approaches", *IEEE Transactions on Automatic Control*, 1989, vol. 34, no. 8, pp. 802-812.

INDEX

Active set, 30, 36
Al-Naib U, 7
Anderson BDO, 100
Antsaklis PJ, 1
Artificial intelligence, 1
Åström KJ, 1, 3
Asymptotic properties, 57
Asymptotic specifications, 63
Backward chaining, 52
Bandwidth, 93
Becker RG, 7
Bezout Identity, 11
Bhattacharyya SP, 7
Birdwell JD, 4, 85
Boyd SP, 6, 7, 9, 19, 23, 24, 27, 29, 32, 82, 97, 98, 107
Boyle JM, 4
Braae M, 67
Buchanan BG, 3
Businger PA, 39
CACSD package, 25, 41
Characteristic Loci, 90
Closed loop performance, 11, 57
Closed loop poles, 22
Closed loop response, 11
    frequency, 28
    step, 28
Closed loop transfer functions, 26
Command history, 43
Command line interpreter, 58, 59
Commands
    ASYM, 69
    COMPLETE, 48, 56, 63, 78
    EDIT, 45, 69
    GROUP PLOT, 59, 72
    HELP, 42, 59
    HELP NVARS, 77
    HELP QSTEP, 77
    NEXT STEP, 42, 45, 56, 59, 68
    NVARS, 78
    OPT, 59, 72
    QSTEP, 78
    SAVE, 80
    SOLVE, 50, 55, 71
    SUGGEST, 42, 46, 56, 59, 73
    SUGGEST EDIT, 47, 61
    SUGGEST OPT, 47, 61
    SVD, 49, 74
Complexity management, 3
Constraints
    active, 64
    asymptotic, 65
    conflict resolution, 66
    conflicts, 44, 63
    frequency domain, 31
    implicit, 49
    linear, 2, 14, 27, 28, 29, 36, 63
    performance, 27, 39, 81, 94
    representation, 30
    satisfied, 64
    time domain, 31
    translation of, 27
Controller
    estimation, 40, 94, 100
    formulae for, 11, 97
    gain, 86
    implementation, 97
    initial stabilizing, 11, 97
    one parameter, 11
    order, 97
    stable initial, 12
    structure, 49

tuning, 99
two parameter, 10, 86
Convexity, 7
Cost function, 2, 26, 29, 81
Database
    facilities for, 53
    FEATURE_DB, 56, 60
    PARAM_DB, 55
    RESP_DB, 53
    SOLVE_DB, 55
    SPEC_DB, 53
DC gain correction, 103
Debugging facilities, 3
Decision variables, 14, 18
Decoupling, 81
Degrees of freedom, 42
Denham MJ, 1
Denominator matrix, 17
Design analysis, 40
Design attributes, 44
Design cycle, 41
Design features, 56, 59
Design specifications, 44, 53
    storage of, 53
Design status, 44
Design tradeoffs, 86
Desoer CA, 6, 12, 13, 15
Direct Nyquist Array, 89
Disturbance rejection, 86, 88, 92
Doyle JC, 48
Dreyfus HL, 1
Dreyfus SE, 1
Edmunds JM, 3
Education, 85
Engineering judgement, 4
Example
    constraint representation, 34, 35
    controller estimation, 103
    controller implementation, 101
    diagonal factorization, 17
    flotation plant, 83
    gyroscope, 80
    level control, 85
    mine mill, 67, 101
    non-minimum phase, 87
    QSTEP, 20
Exception handling, 50

Expert system
    architecture, 3
    external program interface, 52
    structure, 58
Expert system shell, 52
    selection of, 51
Expert systems
    real time, 1
EXPLAIN, 59, 69
External software
    linking to, 52
Factorization
    diagonal, 13, 15, 17, 23, 27, 81
    left, 26
    left-coprime, 11, 15, 17
    right, 26
    right-coprime, 11
    theory, 10
Fegley KA, 6, 7
FIR, 13, 19, 28, 98
    controller implementation, 101
Fourier transform, 13
Francis BA, 12
Frederick DK, 3, 42
Frequency response
    closed loop, 28
Gain matrix, 99
Genesereth MR, 43
Gershgorin circles, 39
Gill-Murray algorithm, 30, 36
Golub GH, 37, 39
Graham N, 1
Graphics, 39
Gustafson CL, 6, 12, 13, 15
Haest M, 1
Hayes-Roth F, 3
Help system, 43
Herget CJ, 3
Horn BKP, 44, 56
Householder transformation, 37
Hulbert DG, 67
Impulse response, 27
Interaction, 88, 92
Interactive design, 4
Interface
    expert system, 25
    to CACSD package, 50, 52

Index.

user, 42
Inverse Nyquist Array, 89
Jackson P, 43
James JR, 3, 51
Jamshidi M, 3
Kailath T, 16
Kimura H, 12, 87
Knowledge base, 3, 52
Knowledge representation, 51
Lagrange multipliers, 37, 44, 65, 75
Larsson JE, 1, 4, 42, 43
Learning, 42
Least squares, 100
Lefrere P, 43
Limebeer DJN, 80, 83
Linear programming, 7
Linear time-invariant, 10
Little JN, 4
Liu Y, 100
LQG/LTR, 4
MacFarlane AGJ, 2, 4, 42, 51
Maciejowski JM, 80, 83, 88
Matrix fractions, 11
    computation of, 12
    coprime, 12
Michie D, 3
Modelling errors, 57
Moore KL, 7
Multivariable systems, 9
MV-CXS, 88, 91
Nett CN, 12
Neural networks, 1
Nolan PJ, 4
Non-monotonic logic, 51
NVARS parameter, 66
Nyquist array, 39
Nyquist plot, 86, 88
Observer-based design, 4
Optimization, 29, 47, 57, 61, 87
Pang GKH, 2, 3, 4, 42
Parameter vector, 26
Parameterization, 18
    first order, 19
    second order, 19
Pattern matching, 54
Performance measures, 92
Persson P, 1, 4, 42, 43

PI control, 85
PI controller, 88
Polak E, 7
Poles
    closed loop, 22
Precompensator, 34
QPSOL algorithm, 36, 37, 44, 64
QR factorization, 36
QSTEP parameter, 18, 28, 66, 81
Quadratic programming, 2, 7, 14, 26, 36
Relative gain array, 88
Remote procedure call, 52, 55
Response
    closed loop, 10
    frequency, 28
    open loop, 49
    step, 28
Reverse frame alignment, 4
Rich E, 1, 2
Robust control, 81
Robustness, 57
Rosenbrock HH, 10
Rychener MD, 3
S domain, 24
Scales LE, 36
Search vector, 26
Sequential loop closing, 88
Single variable design, 88
Singular values, 39, 49
Smith-McMillan form, 12
Specifications, see Design specification
    formulation of, 44
Stable plant, 12
Stable transfer functions, 10
Stein G, 48
Step response, 31
Strictly proper, 6, 10, 97
SV-CXS, 85
Symbolic mathematics, 43
Taylor JH, 3, 42
Tebbutt CD, 19, 29, 30, 42, 52, 83, 85, 95
Trankle TL, 4
Transfer functions
    closed loop, 26
Truncated impulse responses, 101
Unimodular matrices, 10

Unstable plant, 80
Unstable poles, 17
User interface, 42, 51
Users
    experienced, 42
    novice, 42
van Loan CF, 37

Vidyasagar M, 10, 12, 15, 16
Winston PH, 44, 56
Youla DC, 10
Youla parameterization, 11
Zakian V, 7
Zhao Y, 12, 87

Books are to be returned on or before
the last date below.